Advanced Model-Based Engineering
of Embedded Systems

Klaus Pohl · Manfred Broy
Heinrich Daembkes · Harald Hönninger
Editors

Advanced Model-Based Engineering of Embedded Systems

Extensions of the SPES 2020 Methodology

 Springer

Editors
Klaus Pohl
paluno - The Ruhr Institute
  for Software Technology,
University of Duisburg-Essen,
Essen,
Germany

Manfred Broy
Institut für Informatik,
Technische Universität München,
Garching, Bayern
Germany

Heinrich Daembkes,
Airbus Defence and Space,
Ulm,
Germany

Harald Hönninger,
Robert Bosch GmbH,
Renningen, Baden-Württemberg
Germany

ISBN 978-3-319-83876-2   ISBN 978-3-319-48003-9   (eBook)
DOI 10.1007/978-3-319-48003-9

Printed on acid-free paper

This Springer imprint is published by Springer Nature
The registered company is Springer International Publishing AG
The registered company address is: Gewerbestrasse 11, 6330 Cham, Switzerland

Preface

Embedded systems have long become an essential part of our everyday life. They control essential features in our vehicles, such as airbags, braking systems, or power locks, and are used to manage our steadily increasing communication needs by means of Internet routers or cell phones. Embedded systems are essential in application areas where human control is impossible or infeasible, such as adjusting the control surfaces of aircraft or controlling a chemical reaction inside a power plant.

The development of modern embedded systems is becoming increasingly difficult and challenging. Issues that greatly impact their development include the increase in the overall system complexity, their tighter and cross-functional integration, the increasing requirements concerning safety and real-time behavior, the need to reduce development and operation costs, as well as time-to-market.

Need for an integrated, model-based development approach

Many research contributions and development methods aim to address these challenges, and theories for the seamless development of embedded systems have been proposed. However, these solutions address only a small subset of the above-mentioned problems, can only be applied in very specific settings, and lack an appropriate cross-domain validation in representative industrial settings.

The mission of the SPES XT project was thus to focus on the professionalization of a cross-domain, model-based development approach for embedded systems known as the SPES methodology. SPES XT is an joint research project sponsored by the German Federal Ministry of Education and Research. In SPES XT, partners from academia and industry have joined forces in order to enhance a modeling framework based on the latest state of the art in embedded systems engineering to address specific engineering challenges and to ensure that the modeling framework can be applied in embedded systems industries.

Aim of this Book

The aim of this book is to present an overview of the SPES XT modeling framework and to demonstrate its applicability to embedded system development in various representative industry domains. The book explains the basic solution concepts of the SPES XT mod-

Industry challenges, principles, and application

eling framework comprehensively and illustrates the application of these concepts in three major application domains: automation, automotive, and avionics. The book summarizes the lessons learned, outlines evaluation results, and describes how the SPES XT modeling framework can be tailored to meet domain-specific and project-specific needs.

Target Audience

Researchers, practitioners, consultants, and teachers

This book is aimed at professionals and practitioners who deal with the development of embedded systems on a daily basis. They include developers, requirements engineers, software or hardware architects, business analysts, mechatronics experts, safety engineers, testers, and certifiers. The book serves as a compendium for researchers in the field of software engineering and embedded systems, regardless of whether they are working for a research division of a company or are employed with a university or academic research institute. For teachers and consultants, the book provides a solid foundation in the basic relationships and solution concepts for engineering embedded systems and illustrates how these principles and concepts can be applied in practice.

Content of this Book

This book is structured into four parts and sixteen chapters:

Status quo and industry requirements

❏ *Part I — Starting Situation:* This part discusses the status quo of embedded system development and model-based engineering and summarizes the key engineering challenges emerging from industrial practice. Chapter 1 gives detailed insight into the role of embedded systems and outlines the scope of the SPES XT project. Chapter 2 presents two example specifications of embedded systems from the automation and automotive industry. Later on in the book, these case examples will be the main basis for evaluations.

The SPES XT modeling framework

❏ *Part II — Modeling Theory:* This part introduces the backbone of the proposed model-based engineering methodology: the SPES XT modeling framework and its underlying core principles. Chapter 3 presents an overview of the SPES XT modeling framework and introduces its basic methodological concepts. Subsequently, Chapters 4 and 5 place particular emphasis on two major contributions of the SPES XT modeling framework: Chapter 4 introduces a general context modeling framework for embedded systems development and Chapter 5 introduces concepts for the seamless integration of software and systems engineering in industrial development processes.

☐ *Part III — Application of the SPES XT Modeling Framework:* This part describes the application of the SPES XT modeling framework in order to address the major industrial challenges identified. In particular, Chapter 6 proposes solutions to allow early validation of model-based engineering artifacts, Chapter 7 addresses the need to manage the physical context during verification activities, and Chapter 8 details the application of the framework to aid structured engineering of highly interacting system and function networks. Chapter 9 then proposes solutions to support optimized deployments of software, Chapter 10 discusses the opportunities to support modular safety assurance by applying model-based engineering techniques, and Chapter 11 shows the application of the SPES XT modeling framework for systematic variant management and strategic reuse.

Solving major engineering challenges

☐ *Part IV — Evaluation and Technology Transfer:* This part assesses the impact of the SPES XT modeling framework. Chapter 12 summarizes the key lessons learned from exemplary applications in the automation domain. Chapter 13 discusses the value of the SPES XT modeling framework for driving technology transfer. Chapters 14 and 15 provide further evidence for the applicability of the proposed methods by means of industrial tool support as well as the applicability of the methods to industrial case examples. Chapter 16 gives a brief outlook for the future.

Experiences and evidence from industry applications

For further reading, a list of relevant, advanced literature providing deeper insights is given at the end of each chapter.

Acknowledgements

There are many people who have contributed significantly to this book.

Firstly, we would like to thank the members of the Steering Committee of the SPES XT project for their guidance and support throughout the entire project and for encouraging us to document the project results in this book.

Secondly, we would like to thank Ottmar Bender, Dr. Wolfgang Böhm, Peter Heidl, Dr. Stefan Henkler, Dr. Ulrich Löwen, Dr. Andreas Vogelsang, and Dr. Thorsten Weyer for their relentless effort in integrating the different project activities, for many fruitful discussions and suggestions, and for their critical reviews of project milestones. In the same way as for the preciding book "Model-based Engineering of Embedded Systems – The SPES 2020 Methodology" (Springer 2012) much of the content of this book is the result of their devotion and attention to detail.

Thirdly, we would like to thank each and every author of the individual chapters for their patience in the book-writing process, their willingness to revise their chapters time after time, and their cooperation and help in making this book a consistent and integrated product.

Last but not least, we would like to express our deepest thanks to Dr. Thorsten Weyer (from paluno, University of Duisburg-Essen) and his team members Philipp Bohn, Marian Daun, and Bastian Tenbergen for the excellent management of the overall writing and editing process.

The results presented in this book have been made possible through the funding received from the Federal Ministry of Education and Research (BMBF) of the Federal Republic of Germany under grant number 01IS12005. In particular, we would like to thank Prof. Dr. Wolf-Dieter Lukas, Dr. Erasmus Landvogt, and Ingo Ruhmann (all with the BMBF). In addition, we would like to thank Dr. Michael Weber and Jörg Nordengrün of the German Aerospace Center (DLR) for supporting this project.

Furthermore, we would like to thank Tracey Duffy for her valuable language editing assistance and Ralf Gerstner from Springer for his continuous support in publishing this book.

Klaus Pohl
Manfred Broy
Heinrich Daembkes
Harald Hönninger

Summer 2016

Table of Contents

Part I   Starting Situation   1

1 Advanced Model-Based Engineering of Embedded Systems............................ 3
   1.1   Challenges in Embedded System Development..................................4
   1.2   The SPES Engineering Methodology 5
   1.3   Vision and Mission of SPES XT...6
   1.4   Topics not Addressed 7
   1.5   Key Contributions of the SPES XT Approach 8
   1.6   The Future of Embedded Systems 9
   1.7   References 9

2 Running Examples.. 11
   2.1   Introduction 12
   2.2   Automotive Example: Exterior Lighting and Speed Control..............13
   2.3   Automation Example: Desalination Plant 19
   2.4   Summary 25
   2.5   References 25

Part II   Modeling Theory   27

3 SPES XT Modeling Framework 29
   3.1   Introduction 30
   3.2   Structure of the SPES XT Modeling Framework 31
   3.3   SPES Process Building Block Framework 35
   3.4   Specific Extensions of the SPES XT Modeling Framework...............39
   3.5   Summary 41
   3.6   References 42

4 SPES XT Context Modeling Framework 43
   4.1   Introduction 44
   4.2   The SPES XT Context Modeling Framework...............................46
   4.3   Applying Context Models..54
   4.4   Summary 55
   4.5   References 55

5 SPES XT Systems Engineering Extensions 59
   5.1   Introduction 60
   5.2   Standard Engineering Processes 61
   5.3   Integrating Systems and Software Engineering 62

5.4   Summary...70
5.5   References..71

Part III   Application of the SPES XT Modeling Framework   73

6   Early Validation of Engineering Artifacts 75
    6.1   Introduction...76
    6.2   Supporting Artifacts for Validation 80
    6.3   Validation Techniques..82
    6.4   Summary... 101
    6.5   References... 102

7   Verification of Systems in Physical Contexts 105
    7.1   Introduction.. 106
    7.2   Extensions to the SPES Modeling Framework 107
    7.3   Methodological Building Blocks... 113
    7.4   Summary... 116
    7.5   References... 116

8   System Function Networks.. 119
    8.1   Introduction.. 120
    8.2   Extensions to the SPES Modeling Framework 122
    8.3   Methodological Process Building Blocks.................................. 128
    8.4   Summary... 142
    8.5   References... 143

9   Optimal Deployment 145
    9.1   Introduction.. 146
    9.2   Extensions to the SPES Modeling Framework 151
    9.3   Methodological Process Building Blocks.................................. 154
    9.4   Application to the Automotive Example.................................. 166
    9.5   Summary... 167
    9.6   References... 167

10   Modular Safety Assurance 169
    10.1   Introduction.. 170
    10.2   Integrated Safety Framework 173
    10.3   Methodological Building Blocks... 176
    10.4   Summary... 194
    10.5   References... 195

11   Variant Management and Reuse... 197
    11.1   Introduction.. 198
    11.2   Variability Extension to the SPES Modeling Framework 199
    11.3   Methodological Building Blocks... 208

11.4 Summary 220

11.5 References 221

Part IV Evaluation and Technology Transfer   223

12 Experiences of Application in the Automation Domain............................ 225

    12.1 Introduction 226

    12.2 Today's Process 227

    12.3 Technological Hierarchy...228

    12.4 Applying the SPES Viewpoints in the Automation Domain............230

    12.5 Implication for Engineering Tools Used Today 236

    12.6 Summary 237

    12.7 References 238

13 Technology Transfer Concepts 241

    13.1 Introduction 242

    13.2 Technology Transfer in SPES XT...242

    13.3 Guideline Concepts..244

    13.4 Artifact Quality Assessment Framework 247

    13.5 Summary 249

    13.6 References 249

14 The SPES XT Tool Platform... 251

    14.1 Introduction 252

    14.2 Interoperability and Tool Integration Concepts 252

    14.3 Defining the SPES XT Tool Platform 255

    14.4 Summary 261

    14.5 References 261

15 Evaluation of the SPES XT Modeling Framework 263

    15.1 Introduction 264

    15.2 Evaluation Strategy..265

    15.3 Method Toolkit 267

    15.4 Evaluation Landscape...267

    15.5 Applications of the Evaluation Strategy 269

    15.6 Summary 270

    15.7 References 270

16 Outlook... 273

Appendices   277

    A – Author Index..279

    B – Project Structure 285

    C – Members of the SPES XT Project 289

D – List of Publications 291
E – Index 301

Part I

Starting Situation

Klaus Pohl
Manfred Broy
Heinrich Daembkes
Harald Hönninger

1

Advanced Model-Based Engineering of Embedded Systems

The markets for embedded systems are characterized by high innovation pressure, steadily decreasing times to market, and the omnipresent need to reduce development costs. This trend is accompanied by the necessity of developing innovative products with greater functionality and more features that can be sold to customers. In the joint research project "Software Platform Embedded Systems XT" (SPES XT), a group of 21 partners from industry and academia came together to improve the engineering processes for embedded systems in the automation, automotive and avionic industry. In this chapter we give an introduction to the SPES XT modeling framework supporting the seamless model-based engineering of embedded systems and addressing core challenges in todays embedded systems engineering.

© Springer International Publishing AG 2016   3
K. Pohl et al. (eds.), *Advanced Model-Based Engineering of Embedded Systems*,
DOI 10.1007/978-3-319-48003-9_1

1.1   Challenges in Embedded System Development

Embedded systems consist of hardware and software. The computers are often realized by microcontrollers which are typically connected to the whole system of sensors, actors, operator controls, and communication devices. Programs executed by these controllers, known as embedded software, represent an essential part of the systems because they realize the functionality of the systems. Embedded systems are an essential driver for innovation in many domains — for example, automotive, avionics, and industry automation but also in the energy domain and in healthcare, rail, and robotics applications.

Market for embedded systems
In 2012 ninety-eight percent of the microcontrollers produced worldwide are employed in embedded systems. The overall market for embedded systems has increased continually over the last 20 years and is still increasing significantly. The German Federal Association for Information Technology, Telecommunications and New Media (BITKOM) reports that the German market for embedded systems already passed the 20 billion euro mark in 2012.

Market characteristics
Furthermore, the markets for embedded systems are characterized by high innovation pressure, steadily decreasing times to market, and the omnipresent need to reduce development costs. These features are accompanied by the necessity of developing innovative products with greater functionality and more features that can be sold to customers.

Challenges
The result of these demands is that embedded systems are more complex and overall, more embedded systems are being used in contemporary products. The challenge for the industry is to develop these more complex systems with all the required features and to a high level of quality. Mastering these challenges is essential to stay competitive in the markets with respect to innovation, time to market, and cost structures. Failing to meet these goals weakens the competiveness of companies and entire application domains by limiting the innovation potential. Errors and weaknesses in the engineering of such embedded systems may have direct adverse financial consequences or even threaten human physical integrity.

1.2   The SPES Engineering Methodology

In the SPES initiative, the above-mentioned challenges were addressed by proposing a seamlessly integrated model-based, cross-domain engineering approach which addresses the specific concerns of embedded systems and their development processes. A model-based, tool-supported approach based on a solid mathematical foundation as proposed by SPES enables efficient development of embedded systems. The approach starts with an analysis of the system's context and initial customer requirements. This is followed by the gathering of system requirements and specification of the system, architecture design and implementation, and finally, verification and certification of the system.

SPES contributes to meeting the challenges by providing a modeling framework which comprises the fundamental modeling concepts that are needed — including relationships between models, such as refinement — as well as methods for defining mappings between modeling concepts. The SPES modeling framework categorizes these artifacts into four viewpoints and allows artifacts to be specified on a dynamic model of degrees of granularity.

SPES contributes a modeling framework

The viewpoints — which form the horizontal dimension — are based on the *principle of separation of concerns,* which supports a number of views of a system which each serve specific purposes. The aim is to reduce the system's complexity by considering only that information of the system under development which is relevant according to one particular development view.

The SPES modeling framework distinguishes between the following four viewpoints:

Viewpoints

- ❏ Requirements
- ❏ Functional
- ❏ Logical
- ❏ Technical

The different degrees of granularity of systems — which constitute the vertical dimension are concerned with the decomposition of the system into subsystems following the *principle of divide and conquer.* Following this principle, the system is decomposed into smaller and supposedly less complex parts.

Degrees of granularity

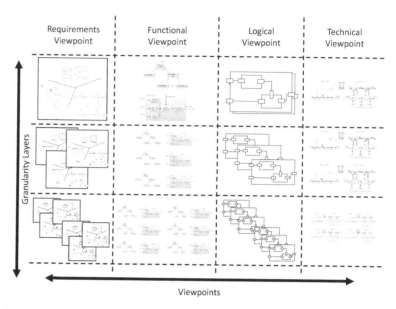

Fig. 1-1   *Structure of the SPES modeling framework*

Practical relevance   Within the SPES XT project, specific artifacts that practitioners produce in their everyday development work were identified and incorporated into the comprehensive SPES XT modeling framework, extending the SPES modeling framework. The artifacts in the framework are separated into those pertaining to the problem space and those pertaining to the solution space. SPES provides specific relationships between these artifacts to facilitate development across different degrees of granularity.

Adaptability   Note that the SPES modeling framework does not prescribe specific modeling techniques or tools. The artifacts can be created and documented using conventional modeling techniques, languages, and tools that most practitioners are already familiar with. In summary, the SPES modeling framework has laid the foundation for engineering for embedded systems in a field that is becoming increasingly demanding.

1.3   Vision and Mission of SPES XT

In the national joint research project SPES XT, a group of 21 partners from industry and academia came together to further extend the seamless methodology and analysis techniques for embedded systems. The research interests focused on six challenges in the en-

gineering of embedded systems identified by the industry partners of
the project as being highly relevant. The following research areas
were addressed by the SPES XT engineering challenges:

❑ Improving the integration between software engineering and
  systems engineering
❑ The integration of systems and ensuring adequate function in
  physical system contexts with a specific focus on system aspects
❑ Design space exploration and optimal deployment of software
  onto hardware components
❑ System quality assurance through validation at the earliest
  possible stage during development
❑ Support for management of system variants and reuse
❑ System qualities (e.g., safety, real time)

The project was set up in a way that allowed a seamless integration *Project characteristics*
of the results across the engineering challenges and across applica-
tion domains, with domain-specific scenarios being discussed within
the engineering challenges. The results were applied to use cases
from exemplary high-tech domains (automotive, avionics, and automation).

The SPES XT project also focused on preparing the results *Industrial acceptance*
achieved in a way that they can be deployed in industrial practice:
the first priority here was to ensure acceptance by developers and
take the current domain-specific development processes into account. To achieve this goal, the project proposed a series of building
blocks that can be tailored for domains, companies, and development teams.

1.4   Topics not Addressed

SPES XT concentrates on key issues in model-based systems engineering. Since the field is so broad, a number of widely known and
very important topics could not be included, for example:

❑ Human-centric engineering
❑ Open context and context uncertainty
❑ Autonomous behavior of systems
❑ Collaboration of embedded systems in dynamic networks
❑ Uncertainty in analysis results, system behavior at runtime, and
  context configurations at runtime
❑ Runtime validation and verification

Foundation for extensions

However, the SPES projects (i.e., SPES 2020 and SPES XT) have developed a broad methodological support and thus laid a foundation for the engineering of embedded systems that can be extended in a variety of directions. A number of the topics listed above, such as open context and collaboration of embedded systems, will be addressed in our future research activities.

1.5　Key Contributions of the SPES XT Approach

As discussed above, the SPES XT project has extended the seamless methodological and technical tool approach for coping with specific challenges in the engineering process for embedded software.

We introduced a set of crosscutting topics which were relevant for all engineering challenges in order to support consistency and the practical applicability of the developed results and to make sure that the following central principles of the SPES methodology were considered in all project results:

❑ Seamless integration of the methodology
❑ Assessment of artifact quality
❑ Integrated tool platform and tool support
❑ Practical applicability

The SPES XT modeling framework is central to the project results. It defines basic terms, concepts, and theory and serves as the basis for all work in the engineering challenges. The framework has been enhanced by specific models and concepts from the engineering challenges.

Introduction of building blocks

To enable seamless methodological support on the one hand and to consider domain-specific requirements on the other hand, SPES XT has developed a set of building blocks that solve the problems of the engineering challenges. These building blocks can be used as elements that constitute domain-specific engineering processes.

Tool support

The integrated tool platform is an enhancement of the reference technology platform from the CESAR[1] project and serves as a framework for integrating academic and industrial tools. From a concept perspective, the tool platform is based on the SPES XT

---

[1] CESAR: Cost-efficient methods and processes for safety relevant embedded systems, ARTEMIS project

modeling framework and allows the engineering challenges to be treated holistically. The platform works in a modular way that enables the use of a suitable subset of platform-based tools which are appropriate for a given problem. These tools will immediately benefit from the additional value provided by the platform, including seamless model-based engineering facilitated by the platform-enabled integration of engineering tools.

For the transfer of the results, such as theories, models, and tools into industrial practice, we have developed guidelines for the practical implementation of the methodology. These guidelines are essential for applying the SPES methodology in specific use cases.

Guidelines for practical implementation

1.6 The Future of Embedded Systems

Embedded systems have undergone and will continue to undergo dramatic change due to the fast pace of development of digital technology. Key areas reflecting that change and the extension of embedded systems in terms of abilities and power include:

❑ Convergence of information and embedded systems
❑ Autonomous systems
❑ Cyber-physical systems
❑ Digital world
❑ Human-centric engineering
❑ Service orientation of embedded systems
❑ Structural and functional integration
❑ Integration with classical information and communication technology

These topics bring in new methodological challenges. Many of these challenges can be addressed by the techniques and principles developed in the SPES projects.

1.7 References

[Pohl et al. 2012] K. Pohl, H. Hönninger, R. Achatz, M. Broyc (Eds.): Model-Based Engineering of Embedded Systems: The SPES 2020 Methodology. Springer, Heidelberg/New York, 2012.

[Rajan and Wahl 2013] A. Rajan, T. Wahl: CESAR – Cost-efficient Methods and Processes for Safety-relevant Embedded Systems. Springer Vienna, 2013.

Frank Houdek
Ulrich Löwen
Jan Christoph Wehrstedt

2

Running Examples

In the following chapters, we introduce advanced concepts that improve the development of embedded systems. To illustrate the concepts and increase comprehensibility, we use examples to demonstrate how the concepts can be applied.

In this chapter, we introduce two illustrative, non-trivial examples that contain typical characteristics of current embedded systems as they are developed in the three SPES application domains, namely automotive, automation, and avionics. In this chapter, we provide an overview of the examples and then in the subsequent chapters, we use specific parts or aspects of the examples to illustrate the SPES methodology.

© Springer International Publishing AG 2016   11
K. Pohl et al. (eds.), *Advanced Model-Based Engineering of Embedded Systems*,
DOI 10.1007/978-3-319-48003-9_2

2.1   Introduction

Two non-trivial real-world examples

Examples are essential when illustrating new concepts. They help to explain abstract ideas as the ideas become more comprehensive. An integrated example helps us to understand how the SPES methodology addresses various aspects of the development challenge and how they interact.

Within this book, we focus on two non-trivial real-world examples to illustrate our concepts. The examples are drawn from the automation and automotive application domains. The first example is an automotive system cluster that contains two typical automotive systems, namely an adaptive exterior lighting system and a speed control system. The main functions of the exterior lighting system are:

❑ Turn signal (direction indicator, hazard warning light)
❑ Low-beam headlights (including daytime running light and cornering light)
❑ High-beam headlights (including automatic high beam if no oncoming vehicle is detected)

The speed control system includes the following functions:

❑ Cruise control and adaptive cruise control
❑ Distance warning
❑ Braking assistant and emergency brake assistant
❑ Speed limiter, including speed sign detection

The second example is a desalination plant with the following main functions:

❑ Pumping water through the processing stations
❑ Reverse osmosis
❑ Controlling water quality

Characteristics of the examples

The use of complex examples also gives an insight into the scalability of our concepts and illustrates how they can be applied to real-world problems. The examples have characteristics which are typical of today's embedded systems:

❑ Highly distributed functionality
❑ Functionality which is partly safety-critical
❑ Real-time demands

- ❑ Software-intense systems (i.e., software is the key element for implementing functionality)
- ❑ Combination of mechanical, mechatronic, electrical, and electronic components
- ❑ Both reactive and controlling behavior

Additionally, the automotive example covers characteristics such as:

- ❑ Many variants of one product in place (e.g., due to different product architecture, local regulations, extra equipment)
- ❑ Product evolution over time; not all engineering and documentation activities (such as the safety case) have to be repeated for the entire product

The desalination plant example from the automation domain covers characteristics such as:

- ❑ Each plant is unique and designed according to the requirements of a specific customer
- ❑ Functionality is greatly determined by software but nevertheless, the core functionality is driven by mechanics; this results in critical dependencies between the mechanics, electrics, and software which have to be managed during engineering

Given the complexity of the examples, it is not feasible to present all of the details within this chapter. Instead, here we give a brief overview of the two examples and elaborate on some aspects that are used later on to illustrate the SPES methodology in more detail.

Disclaimer: The running examples presented in this chapter are inspired by or derived from real-world systems. Nevertheless, we have modified the original real-world systems so that the examples do not describe current or past real-world systems of Daimler AG or Siemens AG.

Disclaimer

2.2 Automotive Example: Exterior Lighting and Speed Control

2.2.1 Project Setting

The project setting of this example is as follows: the automotive original equipment manufacturer (OEM) specifies the requirements of the system in text form at the granularity level *vehicle*, meaning that the system is treated as a black box. In addition, the OEM defines the electric/electronics architecture (E/E architecture) of the vehicle — that is, the electronic control units (ECUs), the communication networks used along with gateways, and the locations of the sensors and actuators. The OEM also defines the basic configuration and extra equipment, which in turn determines optional elements in the E/E architecture and the functional requirements. For safety-related functions, the OEM also conducts a risk analysis to determine which functions are subject to ISO 26262.

Typically, a new project (e.g., for a new vehicle model) will be based on previous projects. New functions are added and existing functions are modified or removed. The automotive example addresses this facet of system evolution by providing four versions of the system specification that build on one another. Therefore, there is a need for a modular safety assessment (see Chapter 10).

Model-based systems engineering is performed primarily by the system provider. The example also covers these activities. However, due to the size of the example, the models shown in this chapter are excerpts and do not represent the complete system.

2.2.2 Functionality

The automotive system cluster contains two automotive systems, namely an exterior lighting system and a speed control system. The main motivation for using two systems is to incorporate some kind of feature interaction: the automatic, speed-dependent high beam benefits from information about the target speed determined by the adaptive cruise control.

Main functions of the exterior lighting system

The main functions of the exterior lighting system are:

❑ **Turn signal (direction indicator, hazard warning light):** In addition to a normal turn signal, the system also offers tip-blinking (i.e., if the pitman arm is used for less than 0.5 seconds, three flash cycles are initiated). In the USA and Canada, an activated daytime running light must be dimmed by 50% during blinking. In order to save energy, the hazard warning light switches its light-dark ratio from 1:1 to 1:2 if the ignition key is removed.

❑ **Low-beam headlights (including daytime running light and cornering light):** The low-beam headlights can be activated manually or — if a rain/light sensor is available — by the environmental light. The cornering light is activated during direction indication if the vehicle is slower than 10 km/h. Ambient light is activated for 30 seconds after a vehicle door has been opened or closed. Starting the engine deactivates the ambient light.

❑ **High-beam headlights (including automatic high beam if no oncoming vehicle is detected):** In automatic mode, the high beam is activated if the camera does not detect an oncoming vehicle. The high-beam headlight illumination area is dependent on the vehicle speed or — if adaptive cruise control is available — on the target speed as determined by the adaptive cruise control system.

An additional optional element of the exterior lighting system is the darkness switch that suppresses exterior lighting. This is relevant for armored vehicles (e.g., police vehicles observing a suspect's house). Some functions of exterior lighting are considered to be safety-relevant: direction indicating is classified as ASIL-A, low beam as ASIL-B.

The main functions of the speed control system are:

Main functions of the speed control system

❑ **Cruise control and adaptive cruise control, traffic jam assistant:** Cruise control maintains the speed of the vehicle at a defined value. The target speed is adjusted via a lever near the pitman arm. The driver can adjust the speed in 1 km/h increments or — in version 2 higher (see also Tab. 2-1) — also in 10 km/h increments.
The adaptive cruise control also takes the distance to the vehicle in front into consideration, meaning that the target speed is the lower of the defined value (set by the driver) and the speed of the vehicle in front. If the vehicle in front stops, our vehicle also stops 2 m behind the other vehicle. If the traffic jam assistant is active, our vehicle automatically starts again if the vehicle in front starts driving again.

❑ **Distance warning:** This function becomes active (by showing a warning symbol) if the distance to vehicle in front is lower than the speed-dependent safety distance.

❑ **Braking assistant and emergency brake assistant:** The braking assistant becomes active if the brake pedal is pressed beyond a certain level. If this happens, 100% braking force is applied to the wheels. The emergency brake assistant reacts to stationary and moving obstacles and gives warnings or activates the brake dependent on the time to collision.

❑ **Speed limiter, including speed sign detection:** The speed limiter function ensures that the vehicle speed does not exceed the maximum speed defined by the driver. In the same way as in the cruise control system, the defined speed can be adjusted in 1 km/h increments or — in version 2 and higher — also in 10 km/h increments. With optional speed sign detection, the speed limit recognized is used as the defined maximum speed.

As mentioned above, there are four versions of the speed control system that reflect the evolution over time. The differences between the four versions are shown in Tab. 2-1.

Tab. 2-1   *Features in the four versions of the speed control system*

Version 1	Version 2	Version 3	Version 4
Cruise control	+ two-level cruise control lever	+ deceleration due to vehicle in front (= adaptive cruise control)	+ traffic jam assistant (i.e., automatic continuation after a holdup)
	Braking assistant	+ distance warning	+ emergency braking assistant
Speed limiter	+ two-level speed limiter lever	+ traffic sign detection	

Because the speed control functions have access to the brake or accelerator, they are also safety-relevant. From an engineering point of view, the evolution of the functionality over time clearly illustrates the need for a modular safety assessment to reduce the overall effort required. For instance, the difference between versions 1 and 2 within the cruise control function is only the comfort option of two increments (1 and 10 km/h) instead of one (1 km/h).

2.2.3   E/E Architecture and Allocation of Functions

As described in Section 2.2.1, in addition to the technical requirements, the OEM defines the E/E architecture of the new vehicle (see Fig. 2-1). The architecture contains both mandatory and optional elements. Some optional elements (e.g., the central display) depend on the extra equipment ordered for an individual vehicle. Other optional elements depend on the vehicle model. In a luxury model, for instance, a two-body controller architecture is used (body controller front, BC_F and body controller rear, BC_R) whereas in a compact vehicle, only one body controller is used (body controller common, BC_C) and the overhead control panel (OCP) is connected to the infotainment gateway (IGW) instead of the BC_F. The E/E architecture also supports electric drive vehicles.

The functional breakdown of a vehicle function and the allocation of the individual functional contribution to individual ECUs is
a key engineering challenge for an OEM. Fig. 2-2 presents one function allocation for the turn indicator functionality.

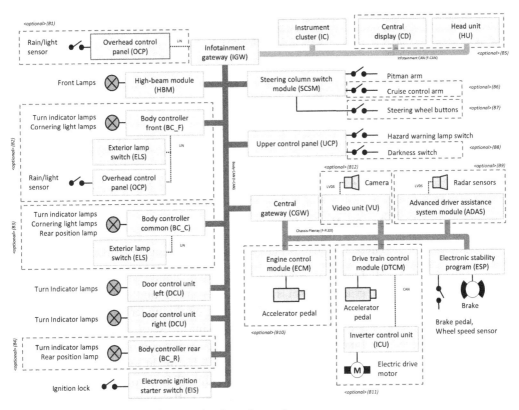

Fig. 2-1   E/E architecture with optional and mandatory elements

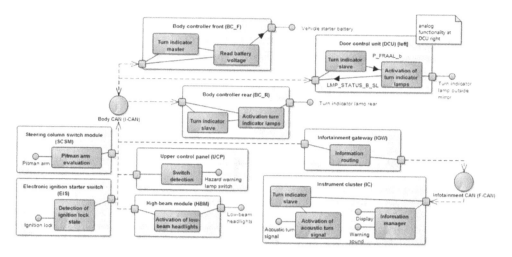

Fig. 2-2   *Function allocation for a turn indicator*

2.2.4   Variability

There are various places where the system contains variability:

❏ Optional equipment: Optional equipment such as (adaptive) cruise control is available dependent on the buyer's choice.

❏ E/E architecture: Depending on the underlying E/E architecture, a component has to communicate with different peers. As the functionality expands (e.g., cruise control vs. adaptive cruise control) the allocation to ECUs might change — for instance, from the engine control module (ECM) to the radar unit (RU).

❏ Country-specific regulations: Different behaviors may have to be implemented dependent on local regulations. In our example, we have the fictitious regulations that (1) in the USA and Canada, while the turn indicator is active the daytime running light must be dimmed by 50%, and (2) that auto-repeat is not permitted when setting the cruise control speed.

❏ Backward compatibility: Starting with version 2, the cruise control and speed limiter can be adjusted in 1 km/h and 10 km/h increments. This requires that the steering column switch module (SCSM) must physically support two increments — that is, the SCSM must support two switching thresholds. Over the production life cycle, it is common to replace a component with its successor (e.g., for cost reasons) while other components remain unchanged. This means that a new ECM that implements the cruise control functionality has to support both one-increment and two-increment SCSMs.

❑ Configuration of behavior: The increment size of 1 and 10 km/h defined in version 2 of the cruise control specification becomes more flexible in version 3. Here, the increments can be configured.

The treatment of variability information in the development process is addressed in Chapter 11.

2.2.5 Engineering Challenges

The main engineering challenges for this example are as follows:

❑ An optimal E/E architecture and distribution of the functionality has to be identified, see Chapter 9.

❑ While functionality or architecture is evolving over time, there is a need to handle safety assessments in a modular way so that (small) changes do not require a full safety reassessment, see Chapter 10.

❑ The product (e.g., ECU) is typically used in many different contexts and has to provide different functionality depending on its environment. Therefore, the engineering process has to explicitly support variability in its many facets, see Chapter 11.

2.3 Automation Example: Desalination Plant

Desalination plants are designed to remove salt from seawater in order to produce drinking water. Desalination typically uses reverse osmosis — a filtration method which requires water to be pumped through membranes at high pressure. This example is based on a real Siemens automation project executed in Spain. The plant has a capacity of 14,000 m³/day and a high level of automation.

Fig. 2-3 shows a schematic overview of this type of desalination plant configuration. Seawater is collected through four beach wells along the coast which are dug into the seashore. From the beach wells, the salt water is pumped through pipelines to the seawater tank, where it is collected, stored, and pre-treated with various chemicals for stabilization and biochemical cleansing before desalination can take place. The subsequent desalination steps include cartridge filtering and repeated treatments until drinking water is obtained in the high-pressure section. Throughout the desalination

Overview of a desalination plant configuration

process, the water quality is consistently monitored and adjusted to ensure that the desalination takes place at optimal efficiency and pollutants and hazardous materials can be removed.

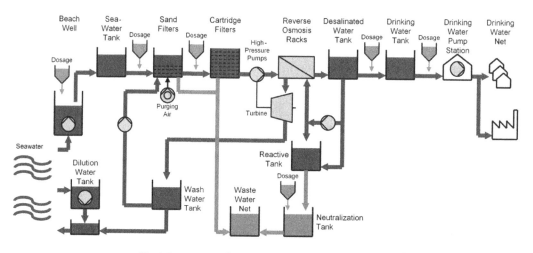

Fig. 2-3   *Configuration of a desalination plant*

To illustrate the engineering challenges, we will focus on the beach well. A beach well has the following general functions:

❑ Collect seawater through subsurface intakes dug into the seashore

❑ Filter seawater through natural sand layers

❑ Pump water from the subsurface intake collection tank to the seawater tank

Adjust the flow into the seawater tank

Automation industry uses piping and installation diagrams

In the automation industry, process plant configurations are designed based on piping and installation diagrams (P&ID). Fig. 2-4 shows the P&ID of one of the four beach wells of our example, outlining the principle parts of the beach wells. Each beach well is equipped with a pump to transport the water collected, a discharge valve through which water is connected to the seawater tank, a bypass control valve which can be used to adjust the flow rate delivered by the pump to avoid damage — for example, due to cavitation — and a hypochlorite valve through which chemicals are added to disinfect the water in order to prevent biological growth in the subsequent filter process.

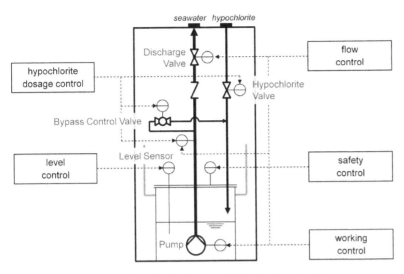

Fig. 2-4   *Piping and installation diagram (P&ID) of a beach well*

The beach wells are equipped with various sensors and actuators to control the desalination process and to avoid damage to the plant — for example, to avoid the beach well pump running dry. Therefore, sensors and actuators are characterized by their interface signals. A small selection of typical signals is shown in Tab. 2-2.

Sensors and actuators control the desalination process

Tab. 2-2   *Sensors and actuators with their interface signals*

	Signal	Type	Direction
Motor	Q_ON	Bool	Automation →Process
	SP_REV	Real	Automation →Process
	FB_ON	Bool	Process →Automation
	FB_POS	Real	Process →Automation
	Error	Bool	Process →Automation
Control valve	SETPOINT	Real	Automation →Process
	FB_POS	Real	Process →Automation
Valve	QON	Bool	Automation →Process
	QOFF	Bool	Automation →Process
	FB_OPEN	Bool	Process →Automation
	FB_CLOSED	Bool	Process →Automation
Level sensor	L	Real	Process →Automation
Flow sensor	F	Real	Process →Automation

Automation software processes signals

These signals must be observed, adjusted, and managed by the automation software so that the beach well can execute its technological functions properly. For example, the set of beach wells must be controlled such that the following requirements are met:

❑ Req. 1: Each pump shall deliver between 400 m³/h and 750 m³/h of water flow.
❑ Req. 2: One pump must always be in standby mode.
❑ Req. 3: A pump shall only run if the fill level of the subsurface tank is sufficient.
❑ Req. 4: The pump loads shall be adapted to achieve and maintain 80% fill level in the pre-treatment seawater tank.
❑ Req. 5: The pumps of the four beach wells start and stop in a cascaded sequence.
❑ Req. 6: The discharge valve must be closed before the pump starts.
❑ Req. 7: The discharge valve must be open after the pump has started.
❑ Req. 8: The discharge valve must be closed after the pump has stopped.
❑ Req. 9: If any pump fails, the standby pump must take over immediately.

In this example, the system under development is the customer-specific automation software and hardware of the desalination plant, whereby the automation hardware is configured based on automation hardware components delivered out of the box by specific suppliers.

Engineering of the automation of a desalination plant is executed within customer projects. This process is driven by a higher level project management process. Different companies and different disciplines (civil, mechanical, electrical, automation engineering) collaborate within such a project, each using their own methods, tools, and perspective on the desalination plant. A common naming system is an important concept for integration.

Main engineering challenges

The main engineering challenges for this example are as follows:

❑ The automation engineering is subordinated to the design decisions concerning the technological process. The parallel work of the different disciplines results in complex interdependencies, manifold change requests between the disciplines, and complex integration activities which have to be managed, see Chapter 11.

❑ Customer-specific requirements result in a large volume of order-specific engineering. To reduce the order-specific engineering efforts required and to build on proven solutions, a major challenge is to establish and improve the reuse of engineering artifacts, see Chapter 11.

❑ Project business entails high commercial and technical risks. Therefore, design decisions have to be validated as early as possible, see Chapter 6.

These general challenges have to be solved in a demanding environment with respect to tools as well as the quantities of input and output data and artifacts.

Managing engineering tools is challenging

For the engineering of desalination plants, a wide variety of tools ranging from specific MS Excel sheets to powerful domain-specific engineering platforms is used by the different disciplines. It is typically a challenge to manage these engineering tools to guarantee a seamless engineering workflow. In our running example, we focus on the following aspects (see also Fig. 2-5):

❑ Requirements are documented informally in textual form.

❑ P&IDs and the required electrical drawings are created by computer-aided engineering (CAE) tools. In our example, we used COMOS, see [Siemens 2015a].

❑ Distributed control systems are used for the engineering for the automation software. These are specific systems developed for automating plants in process industries. In our example, we used SIMATIC PCS 7 (see [Siemens 2015b]).

Typical engineering artifacts

To give an idea of the typical kinds and numbers of engineering artifacts considered for the desalination plant based on the real project, the following is a short overview:

❑ From the perspective of the automation of the desalination plant, the requirement and design specification describes the automation functions in textual form as well as the graphical user interface for the operator of the plant. This document has around 100 pages and is typically supplemented by numerous system specifications for the automation products used.

❑ P&IDs describe the operational context of the automation. This is often one large overview plan describing the technological process, see Fig. 2-4.

❑ Circuit diagrams describe the electrical engineering aspects. Usually, there are around 100 plans, which include, for example, a list of sensors and actuators and the resulting list of signals between the technological process and the automation. In our example, there are around 500 sensors and actuators and therefore around 1000 signals which have to be considered.

❑ The automation software is specified using continuous and sequential functions charts (CFC and SFC respectively). These descriptions are well known in automation and are based on an international standard. For more details, see [Urbas 2012]. In our example, there are around 400 CFCs and 10 SFCs.

❑ Specific tools are used for the design and implementation of the graphical user interface for the operator-specific tools which are usually part of the process control system. In our example, there are around 30 operator pictures.

❑ The specification of the hardware configuration for a process control system depends heavily on the process control system used. Specific tools are provided for specifying the field devices, input and output modules for connecting the signals, the programmable logic controllers (PLCs), and the server and clients for the graphical user interface including the communication infrastructure between these components. In our example, we have one server, around ten PLCs, and numerous field devices as well as input and output modules for connecting around 1000 signals.

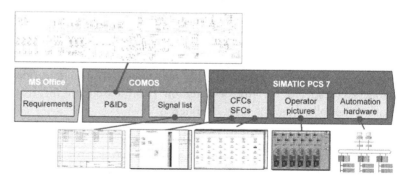

Fig. 2-5   *Engineering tool chain with COMOS and SIMATIC PCS 7*

With respect to the challenges, methods, tools, and systems for automation, this example is a typical representation of an industrial plant in process industries. Besides water technologies, other examples of process industries are energy generation, oil and gas, chemical, pharmaceutical, or food and beverage.

2.4 Summary

This chapter introduced two systems that will be used in the subsequent chapters to illustrate the SPES XT concepts. The examples have not been specifically designed to fit the SPES XT concepts and are instead based on real-world systems along with all their constraints and engineering challenges.

2.5 References

[Siemens 2015a] Siemens AG: COMOS at a glance.
     http://w3.siemens.com/mcms/plant-engineering-software/en/comos-
     overview/Pages/Default.aspx. (Accessed January 15, 2015).

[Siemens 2015b] Siemens AG: SIMATIC PCS 7.
     http://w3.siemens.com/mcms/process-control-systems/en/distributed-control-
     system-simatic-pcs-7/Pages/distributed-control-system-simatic-pcs-7.aspx. (Accessed January 15, 2015).

[Urbas 2012] L. Urbas: Process Control Systems Engineering. Oldenbourg Industrieverlag, Munich, 2012.

Part II

Modeling Theory

Wolfgang Böhm
Marian Daun
Vasileios Koutsoumpas
Andreas Vogelsang
Thorsten Weyer

3

SPES XT Modeling Framework

As embedded systems evolve into more and more complex structures to meet the continuously increasing complexity of requirements, they face a variety of challenges. In particular, the involvement of multiple engineering disciplines targeting cross-cutting aspects of the system under development makes the situation even more challenging. Hence, there is a great need to establish a seamless modeling framework that on the one hand, facilitates reuse and automation, while on the other hand, is independent of any application domain. The modeling framework must provide appropriate models and description techniques for modeling the different aspects and artifacts of system development as well as methods and process techniques for creating and analyzing such artifacts. Therefore, this chapter introduces the SPES XT modeling framework, which aims to address these issues.

© Springer International Publishing AG 2016   29
K. Pohl et al. (eds.), *Advanced Model-Based Engineering of Embedded Systems*,
DOI 10.1007/978-3-319-48003-9_3

3.1   Introduction

The SPES modeling framework

From 2010 to 2012, the SPES modeling framework [Broy et al. 2012] was developed in a close collaboration between academia and industry. The SPES modeling framework, which is a structured collection of modeling concepts, enables the seamless model-based engineering of embedded software and relies on the core principles of *divide and conquer* and *separation of concerns*. The framework allows us to manage the complexity of modern embedded systems during the software engineering process. Furthermore, it allows us to apply formal methods for verification and validation purposes, which in turn, for instance, fulfills the need for safety-critical embedded software to work correctly.

SPES XT extensions

The original SPES modeling framework already emphasized the use of the framework for documenting and analyzing certain quality aspects such as safety (see [Höfig et al. 2012]) and real time (see [Hilbrich et al. 2012]). However, in order to address the specific engineering challenges identified, the SPES modeling framework has been extended in two different directions:

❑   *Core methodological extensions* to address additional methodological aspects of a general modeling theory for embedded systems

❑   *Specific methodological extensions* to address specific engineering challenges in the engineering of embedded systems

The SPES XT modeling framework comprises the SPES modeling framework and, among others, three *core methodological extensions* to the original SPES modeling framework described in [Broy et al. 2012]:

❑   The *SPES XT Process Building Block Framework* allows the definition of customized engineering processes for specific purposes based on the artifacts defined in the SPES XT modeling framework (Section 3.3).

❑   The *SPES XT Context Modeling Framework* allows consistent documentation and analysis of properties or assumptions about the context of embedded systems and software (Chapter 4).

❑   The *SPES XT Systems Engineering Extensions* allow the SPES XT modeling framework to be applied within the overall systems engineering process for embedded systems including the

different systems engineering disciplines, like electrical engineering, mechanical engineering as well as software engineering (Chapter 5).

The *specific methodological extensions* of the SPES XT modeling
framework comprise extensions to the original framework with
respect to the following specific challenges in the engineering of
embedded systems:

- ❑ *Early validation of engineering artifacts (Chapter 6)*
- ❑ *Verification of systems in physical contexts (Chapter 7)*
- ❑ *System function networks (Chapter 8)*
- ❑ *Optimal deployment (Chapter 9)*
- ❑ *Modular safety assurance (Chapter 10)*
- ❑ *Variant management and reuse (Chapter 11)*

3.2   Structure of the SPES XT Modeling Framework

In the SPES XT modeling framework, a system is described from
different *viewpoints* capturing different stakeholder concerns and
with varying *degrees of granularity*. A system description from a
specific viewpoint and with a specific degree of granularity is called
a *view* [ISO/IEC/IEEE 42010].

3.2.1   Viewpoints

The SPES XT modeling framework facilitates the seamless, model-based engineering of embedded software with four core viewpoints. *Four core viewpoints*
These viewpoints distinguish between the problem space and the
solution space and between functional, logical, and technical solution concepts. In doing so, the viewpoints address the concerns of
different stakeholders. We consider these viewpoints and the corresponding concerns as core viewpoints common for embedded software across all application domains. The four core viewpoints and
their concerns are:

The *requirements viewpoint* supports the requirements engineer- *Requirements viewpoint*
ing process in eliciting, documenting, negotiating, and validating
requirements for the system under development (SUD). To distinguish between different types of requirements such as assumptions
and constraints, goals, behavioral requirements, and more detailed
solution-related requirements, the requirements viewpoint differentiates between four types of models: the context model, the goal

model, the scenario model, and the solution-oriented requirements model. For details on the requirements engineering viewpoint, see [Daun et al. 2012].

Functional vimewpoint   The *functional viewpoint* supports the development of a functional system specification for the SUD. Requirements captured in the requirements viewpoint are structured with respect to user functions, which specify patterns of use from an actor's (i.e. an human user or an external system) point of view. User functions are specified by system functions, which are the functional building blocks that the system is intended to provide. These types of models enable a precise analysis of functional dependencies and feature interactions [Vogelsang and Fuhrmann 2013]. For details on the functional viewpoint, see [Vogelsang et al. 2012].

Logical viewpoint   The *logical viewpoint* supports the solution design for the SUD. In this viewpoint, functional building blocks (captured in the functional viewpoint) are realized by communicating components arranged in a component architecture. In contrast to the functional viewpoint, the models of the logical viewpoint are not structured solely with respect to functionality but rather in terms of architectural design. Here, aspects such as the organizational structure, dependability, maintainability, and reusability also play an important role (e.g., components that implement reliability mechanisms). For details on the logical viewpoint, see [Eder et al. 2012].

Technical viewpoint   The *technical viewpoint* supports the technical implementation of the SUD. Components, which are captured in the logical viewpoint and which describe abstract solution elements, are refined into software modules that are executed on a specific execution platform. The mapping from components in the logical viewpoint to elements of the technical viewpoint is called *deployment* (see also Chapter 9). This viewpoint defines how the software and hardware interact to realize the system goals. For details on the technical viewpoint, see [Weber et al. 2012].

3.2.2 Degree of Granularity

The viewpoints describe the SUD with respect to different concerns. However, these descriptions may vary in their degree of granularity. For complex systems in particular, it is reasonable to start with rather high-level descriptions of requirements, functions, components, and technical devices. Once these high-level descriptions have been created, these views are typically refined and detailed step by step. Therefore, the SPES XT modeling framework supports views with different degrees of granularity.

To change the degree of granularity for a given view to a higher degree, a low-degree view is decomposed into a number of more detailed system views following the principle of divide and conquer. This step can be performed from a view of any viewpoint (e.g., a function is decomposed into subfunctions or a component is decomposed into subcomponents). As a result, we get a view for each element resulting from the decomposition. These views have a higher degree of granularity and also a smaller scope compared to the original system element from which they are derived. Furthermore, we can look at these detailed elements again from different viewpoints to specify their requirements, functions, logical components, and technical components. We can visualize this procedure with a tree-based representation as shown in Fig. 3-1. The root node of the tree represents the entire SUD (S). For this system, we document the requirements in the requirements viewpoint (R), specify the functions in the functional viewpoint (F), design a component architecture in the logical viewpoint (L), and describe the intended execution platform in the technical viewpoint (T). When decomposing the system into an architecture of communicating components in the logical viewpoint, we increase the degree of granularity. In the figure, the system S is decomposed into three logical components (LC1-3), each building a separate engineering path with a higher degree of granularity and with a smaller scope. For each of these logical components, we again document the requirements (R), specify the functions (F), design a component architecture (L), and describe the intended execution platform (T).

In this example, we used the view of the logical viewpoint to change the degree of granularity, i.e., the decomposition of the logical viewpoint model determines the next degree of granularity. In general, we can decide separately for each system element which viewpoint to use to increase the degree of granularity. Automotive manufacturers, for example, usually structure their vehicle systems with respect to functions (functional viewpoint), which are subsequently engineered by suppliers (who may specify component architectures for the functions). System integrators, on the other hand, would usually detail views of the technical viewpoint because they integrate hardware, basic software, and application software.

Degree of granularity

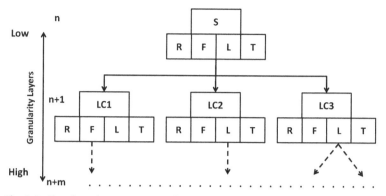

Fig. 3-1   *Tree-based representation of granularity layers*

Engineering path   The development of views for a system element with a specific degree of granularity is summarized in an *engineering path*. When decomposing a system element into a set of more detailed system elements, the engineering process is split up into an engineering path for each detailed system element. We illustrate this idea of engineering paths in Fig. 3-2, which shows a different representation of the example from Fig. 3-1. In the figure, the engineering path for system S is split up into three engineering paths for logical components LC1-3, which together define the next granularity layer. Within the engineering paths, again the models of the viewpoints are created for the corresponding system element.

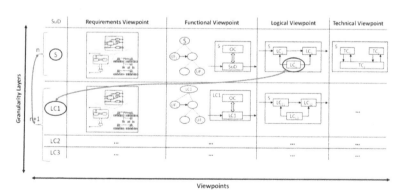

Fig. 3-2   *Structuring of the engineering process by granularity layers*

Development steps   When considering the models that represent the views in the SPES XT modeling framework, we can explain the progression with decomposition and refinement relations:

❑   *Decomposition*: Decomposition describes a model as an ensemble of smaller model elements and their relationships. In the SPES XT modeling framework, decomposition is used to increase the degree of granularity. Functional decomposition, for example, allows a structuring of the interface behavior of the overall system in terms of functions leading to a function hierarchy. A logical decomposition describes the decomposition of the overall system into subsystems.

❑ *Refinement*: Refinement describes the transformation of an abstract model into a more concrete model with the same properties as the abstract model. For instance, the refinement relations can address the description of the system interface. Technical signals are a refinement of logical messages, which may be refinements of events used to formulate requirements. The resulting behavior described by the models must be consistent with each other. For example, the logical model must fulfill the behavior of the functional model but may also ensure some quality characteristics (e.g., behavior in the case of failures).

The use of mechanisms for abstraction, decomposition, or the definition of different degrees of granularity is common in the engineering of embedded systems. The SPES XT modeling framework employs these mechanisms to enable systematic engineering of embedded systems. Furthermore, these mechanisms are mapped to specific model types with clear relationships to each other.

Contributions of the SPES XT modeling framework

3.3   SPES Process Building Block Framework

It is important to distinguish between the artifact structure that is used in system development and the development process itself. On the one hand, the artifact structure defines which contents are worked out and documented and how they are related. On the other hand, the development process defines by which techniques and in which order the artifacts are created. While the SPES XT modeling framework focuses on the artifact structure, the SPES XT Process Building Block Framework [Daun et al. 2016] provides a methodology for documenting processes, which describe the creation and analysis of artifacts. Hence, the SPES XT Process Building Block Framework and the SPES XT modeling framework complement each other and provide a consistent methodology for the engineering of embedded software.

Defining development processes

3.3.1   Overview of the Framework

Building blocks

The relationships between artifacts are specified by relations. Such a relationship can be expressed by a process building block that defines a general technique for artifact creation and analysis. This can be described, for example, as an algorithm, a guideline, or a tool. A building block has an input (e.g., an engineering artifact or stakeholder knowledge) and an output, (e.g., an engineering artifact or an analysis result). Furthermore, preconditions and postconditions can be associated with building blocks to meet custom requirements of applicability for a specific process. There are two main types of process building blocks:

Construction building block

❑   The output of a *construction building block* is always an artifact that belongs to a specific viewpoint. For example, the derivation of a logical architecture from a functional architecture creates an artifact that belongs to the logical viewpoint. A construction building block may or may not have some artifact-related input. For example, a context diagram may be created from scratch based on information that might not even be documented.

Analysis building block

❑   *Analysis building blocks* are used to analyze engineering artifacts. They could be, for example, a consistency check between two artifacts from different viewpoints or from different degrees of granularity. Analysis building blocks have some input and their output is an analysis result, which can be an artifact that cannot be mapped to a specific viewpoint of the SPES XT modeling framework even though it might be a model.

Combining building blocks

Processes can be combined to form larger building blocks. On the one hand, this building block concept provides flexibility as it does not impose a rigid process. On the other hand, it allows building blocks to be reused within the same process as well as in other processes.

Process definition

To define a development process, building blocks are combined: both construction building blocks and analysis building blocks can be used. The output of one building block can be the input of another building block. Furthermore, one building block's postcondition might ensure that another building block's precondition is met. Not only building blocks themselves but also sequences of building blocks can be reused in other processes.

Building block templates

Custom process templates can be defined to characterize common processes in various engineering challenges. For example, the derivation of a functional architecture from a requirements specification based on message sequence charts (MSCs) can be defined as a building block template which can be reused in a variety of engineering challenges.

3.3.2 Relationship to the SPES XT Modeling Framework

An important distinction in the SPES XT building block framework is the structuring of the results of the development process, which are artifacts, and the relationships between them. The way artifacts are defined during the development process often varies. However, there are some basic concepts and relationships defined in the SPES XT modeling framework that can be summarized as follows:

❑ *Init building blocks:* An artifact that belongs to a viewpoint can be created from scratch, meaning that the corresponding technique does not rely on other artifacts. For example, the initial artifacts of the requirements viewpoint are developed from scratch (see Fig. 3-3 (1)). To do so, the requirements engineer typically consults the stakeholders and elicits their intentions.

❑ *Relationships of artifacts between viewpoints:* a construction building block can relate artifacts of viewpoints on the same engineering path. This enables relationships between artifacts of any viewpoint to be modeled on the same degree of granularity (horizontal relation of artifacts, see Fig. 3-3 (2)). This allows, for example, the generation of an initial version of the functional design (i.e., a major engineering artifact in the functional viewpoint) from the existing requirements artifacts.

❑ *Relationships of artifacts between different degrees of granularity:* construction building blocks can relate artifacts with the same viewpoint but with different degrees of granularity. By combining the building blocks, we achieve higher-order relationships between artifacts with different degrees of granularity (vertical relation of artifacts among the viewpoints, see Fig. 3-3 (3)).

❑ *Relationships between artifacts belonging to the viewpoints and artifacts outside the viewpoints:* in all phases of the development process, there are certain artifacts that are defined but are not related to the SPES XT modeling framework, see Fig. 3-3 (4). For example, the logical architecture is often used to create simulations and tests for the SUD. The results of this simulation and testing phase are not part of the SPES XT modeling framework, although they may be used for the correction of the functional design, which is again an artifact belonging to the SPES XT modeling framework. Thus, analysis building blocks are very important, not only because they allow the analysis of certain artifacts but also because they provide the ability to trace back the derivation of certain artifacts.

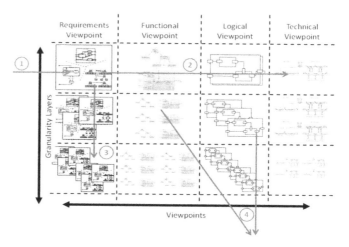

Fig. 3-3   *Relationships between artifacts in the SPES XT modeling framework*

3.3.3   Example: Beach Well Case Study

The following example depicted in Fig. 3-4 illustrates the use of the SPES XT process building block framework for the beach well case study. It describes the process of creating a formal requirements specification from informal system requirements (creation block C-1), a corresponding functional design (creation block C-2), the creation (analysis block A-3) and use (analysis block A-4) of a verification model, and the correction of the functional design (creation block C-5) and requirements specification (creation block C-6) with the selected degree of granularity.

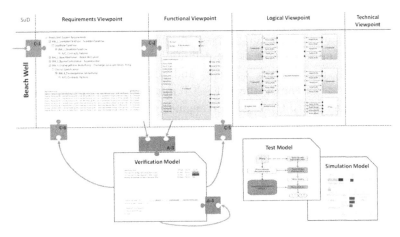

Fig. 3-4   *Artifact structure and building blocks for the beach well case study*

Fig. 3-5 illustrates the combination of the basic building blocks defined above to describe an exemplary development process for beach wells. A template is filled out for each building block, with the applied methodology explained in more detail. Based on the schema above, custom development processes may be defined for the aspects of interest. For example, in the logical viewpoint, a logical architecture may be derived from a functional architecture. Furthermore, the logical architecture may be used for testing and simulation. These results are then needed again for the correction of the functional design, and so on.

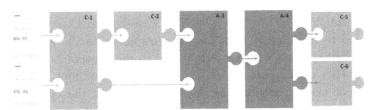

Fig. 3-5   *Building blocks of the engineering process for beach wells*

3.4   Specific Extensions of the SPES XT Modeling Framework

The extensions of the modeling framework support the need for tailoring and instantiating the modeling framework in different problem domains and for specific problem classes. Thus, the SPES XT modeling framework provides different extension mechanisms for specific purposes to allow for seamless artifact-based engineering of embedded systems.

Core model

The main strategy for purpose-specific extensions of the SPES XT modeling framework is the definition of metamodel extensions that partly use the core concepts and add further concepts. This technique of extending the SPES XT modeling framework has already been applied to six general engineering challenges in the engineering of embedded systems. However, the SPES XT modeling framework can be extended in a similar way to deal with other problem classes as well. The SPES XT modeling framework core model, consisting of the four core viewpoints and the core views on different degrees of granularity, can be extended by certain elements. For example, the basic concepts of a functional architecture from the core functional viewpoint can be extended to allow more sophisticated functional architectures to consider collaboration between different networked systems.

Core model extensions

Core model extensions are designed to describe individual extensions for each domain. On the one hand, modeling framework extensions that adapt the core metamodel to the needs of a specific domain may be defined. On the other hand, custom building blocks may be useful for modeling individual development processes according to the needs of a specific engineering challenge. In conclusion, each engineering domain defines its own extensions based on the core model to address its specific engineering challenges.

Crosscutting concerns

For some purposes, it is necessary to address crosscutting concerns, which affect each viewpoint, each degree of granularity, and furthermore, other extensions. Such crosscutting concerns might be, for example, real time or variability. Crosscutting extensions are related to qualities of embedded systems affecting all viewpoints. Crosscutting extensions are orthogonal to the viewpoints (see Fig. 3-6). They define ontological elements and relationships that may enhance existing engineering artifacts. Therefore, the ontological entities may be defined specific to viewpoints, specific to degrees of granularity, or even applicable to all viewpoints, or to all degrees of granularity. The ontological elements of the extensions are related to each other, for example, to ontological elements of another viewpoint or another degree of granularity. The crosscutting specific ontological elements must be related to the ontological elements of the core engineering artifacts of the SPES XT modeling framework. Defining further crosscutting extensions means that the artifacts of two crosscutting extensions will probably be related to one another as well. The SPES XT modeling framework in combination with the SPES XT process building block framework provides the necessary modeling concepts to achieve a unified and consistent method of modeling ontological relationships of crosscutting extensions.-

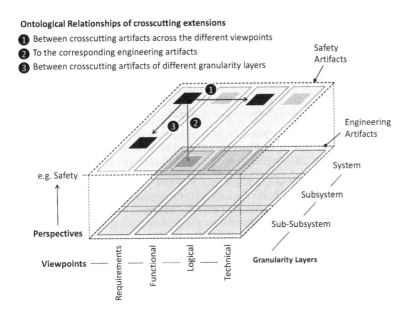

Ontological Relationships of crosscutting extensions
① Between crosscutting artifacts across the different viewpoints
② To the corresponding engineering artifacts
③ Between crosscutting artifacts of different granularity layers

Fig. 3-6   *Defining crosscutting extensions in the SPES XT modeling framework (see [Heuer et al. 2013])*

3.5   **Summary**

In this chapter, we presented a conceptual overview of the SPES XT modeling framework. In particular, we illustrated how the framework supports the specification process for embedded software with different degrees of granularity. Furthermore, we introduced the SPES XT Process Building Block Framework, which provides a means for defining engineering processes based on fine-grained methodological building blocks. Hence, the SPES methodology provides the necessary concepts and techniques for defining artifact structures, their relationships and the corresponding engineering steps. Finally, we presented extension mechanisms of the SPES XT modeling framework. These mechanisms can be used to extend the methodological framework to deal with customized engineering problems from different application domains.

3.6   References

[Broy et al. 2012] M. Broy, W. Damm, S. Henkler, K. Pohl, A. Vogelsang, T. Weyer: Introduction to the SPES Modeling Framework. In: K. Pohl, H. Hönninger, R. Achatz, M. Broy: Model-Based Engineering of Embedded Systems: The SPES 2020 Methodology. Springer, Heidelberg/New York, 2012.

[Daun et al. 2012] M. Daun, B. Tenbergen, T. Weyer: Requirements Engineering Viewpoint. In: K. Pohl, H. Hönninger, R. Achatz, M. Broy: Model-Based Engineering of Embedded Systems: The SPES 2020 Methodology. Springer, Heidelberg/New York, 2012.

[Daun et al. 2016] M. Daun, P. Bohn, J. Brings, T. Weyer: Structured Model-Based Engineering of Long-living Embedded Systems: The SPES Methodological Building Blocks Framework. Softwaretechnik-Trends, Vol. 36, No. 1, 2016.

[Eder et al. 2012] S. Eder, J. Mund, A. Vogelsang: Logical Viewpoint. In: K. Pohl, H. Hönninger, R. Achatz, M. Broy: Model-Based Engineering of Embedded Systems: The SPES 2020 Methodology. Springer, Heidelberg/New York, 2012.

[Heuer et al. 2013] A. Heuer, T. Kaufmann, T. Weyer: Extending an IEEE 42010-Compliant Viewpoint-Based Engineering Framework for Embedded Systems to Support Variant Management. In: Embedded Systems: Design, Analysis and Verification, IFIP Advances in Information and Communication Technology, Vol. 403, Springer, Heidelberg 2013, 283-292.

[Hilbrich et al. 2012] R. Hilbrich, J. van Kampenhout, M. Daun, T. Weyer, D. Sojer: Modeling Quality Aspects: Real-Time. In: K. Pohl, H. Hönninger, R. Achatz, M. Broy: Model-Based Engineering of Embedded Systems: The SPES 2020 Methodology. Springer, Heidelberg/New York, 2012.

[Höfig et al. 2012] K. Höfig, M. Trapp, B. Zimmer, P. Liggesmeyer: Modeling Quality Aspects: Safety. In: K. Pohl, H. Hönninger, R. Achatz, M. Broy: Model-Based Engineering of Embedded Systems: The SPES 2020 Methodology. Springer, Heidelberg/New York, 2012.

[ISO/IEC/IEEE 42010] ISO/IEC/IEEE. Systems and software engineering – Architecture description. ISO/IEC/IEEE 42010:2011(E), International Organization for Standardization, Geneva, Switzerland. 2011.

[Vogelsang and Fuhrmann 2013] A. Vogelsang, S. Fuhrmann: Why feature dependencies challenge the requirements engineering of automotive systems: An empirical study. In: Proceedings of the 21st IEEE International Requirements Engineering Conference (RE). 2013.

[Vogelsang et al. 2012] A. Vogelsang, S. Eder, M. Feilkas, D. Ratiu: Functional Viewpoint. In: K. Pohl, H. Hönninger, R. Achatz, M. Broy: Model-Based Engineering of Embedded Systems: The SPES 2020 Methodology. Springer, Heidelberg/New York, 2012.

[Weber et al. 2012] R. Weber, P. Reinkemeier, S. Henkler, I. Stierand: Technical Viewpoint. In: K. Pohl, H. Hönninger, R. Achatz, M. Broy: Model-Based Engineering of Embedded Systems: The SPES 2020 Methodology. Springer, Heidelberg/New York, 2012.

Marian Daun
Bastian Tenbergen
Jennifer Brings
Thorsten Weyer

4

SPES XT Context Modeling Framework

When developing embedded systems, the context is of vital importance as embedded systems interact with the context through sensing and actuation. There is a multitude of context information, which is relevant for embedded systems engineering: on the one hand, external systems and human users interacting with the system under development constrain the specific interaction among them. On the other hand, properties of these external systems and human users and also laws, regulations, or standards constrain the the system under development. In this chapter, we provide a context modeling framework that seamlessly integrates into the SPES XT modeling framework and takes different viewpoints, perspectives, and degrees of granularity into account. This chapter presents:

☐ The context of knowledge model, which allows the documentation of knowledge sources, the context information they provide, and their mutual influences
☐ Three operational context models, which allow the documentation of static-structural and functional properties as well as the context's behavior
☐ The relationship between the different context models and other SPES modeling framework artifacts with regard to specific application scenarios

© Springer International Publishing AG 2016
K. Pohl et al. (eds.), *Advanced Model-Based Engineering of Embedded Systems*,
DOI 10.1007/978-3-319-48003-9_4

4.1   Introduction

The importance of context information

Software-intensive embedded systems differ from information systems in that they interact with their operational context through sensors and actuators. Therefore, the context of such systems must be carefully considered during their development. Without proper consideration of the system's context, functional suitability, performance, or system safety may be impaired. Consider the automotive system cluster from Chapter 2. In the event of an emergency brake situation, we want the light control unit to flash the hazard warning light — from the perspective of the braking system, the light control unit is in its context and vice versa. Specifically, it is important for the control unit to handle the information correctly, that is, by flashing the lights only given a specific speed and deceleration rate, and to do so immediately upon impact in the event of a crash. However, the system's context pertains not only to its behavior during operation. Context information may also constrain the development itself. For example, laws in different countries may prohibit or demand system functionality, such as with different laws pertaining to the maximum permissible tint in vehicle windows in the United States vs. Germany requiring automated window tinting systems to limit the tint (see the German road regulations [StVZO 2013], and the Official Code of Georgia [Georgia 2013]). This type of context information therefore constrains both the system as well as its development.

4.1.1   Advantages of Explicit Consideration of Context

Conflict resolution and assumption validation

On the one hand, context information can help the developers in eliciting necessary and relevant information for system development. On the other hand, the context is essential to identify conflicts between assumptions and facts in the system's specification or conflicts between stakeholder intentions. Specifically, the benefits of explicit documentation of context information include:

❑ *Separation between context and system* allows us to ascertain what aspects of the system are subject to engineering activities and what aspects constrain the system and/or its engineering process

❑ *Identification of information sources* within the context allows us to determine the intended usage context of the system [Daun et al. 2012], which is a prerequisite for requirements management activities — for example, requirements prioritization

☐ *Documenting context information* is a prerequisite for making knowledge about the system's intended and future use persistent and available at all times to all involved personnel, thereby counteracting the knowledge drain due to outsourcing, for example

☐ *Support for decision making* in the case of conflicts between stakeholders, interface specifications, or invalid development assumptions

☐ *Support for change management* as changes in the context of a system may affect its suitability, functionality, or safety [ISO/IEC 25010] if they are not properly incorporated into the system

4.1.2   Existing Context Modeling Concepts

In requirements engineering in particular, the need to explicitly document implicit knowledge about the problem domain is vitally important [Gause 2005] and various approaches have been suggested. In goal-oriented approaches (e.g., [Van Lamsweerde 2010], [Yu 1995]) top-level goals are refined into specific requirements with respect to context information. This makes the fulfillment of the goal dependent on the system itself, external systems, or human users.

Ontology-based context modeling approaches have also been proposed in the past (e.g., [Strang et al. 2003], [Bergh and Coninx 2006]). These approaches support the documentation of state-based behavior. This is a prerequisite for various quality assurance and analysis approaches such as model checking of static properties of development artifacts (e.g., [Dhaussy et al. 2009]) as well as impact analysis of context changes (e.g., [Alfaro and Henzinger 2001]). However, a prerequisite for explicit documentation in the first place, is a proper understanding of the problem domain. For example, in [Zave and Jackson 1997], [Jackson 2006], and [Jin and Liu 2006], particular emphasis is placed on the distinction between the system and the system's context.

4.2   The SPES XT Context Modeling Framework

Separation of concerns   When considering the context, we first have to adequately separate the relevant aspects of the context from the irrelevant parts. In addition, it is important to distinguish between context information, which is immutable and cannot be influenced through development activities and parts of the system, which can be changed through development. Therefore, each piece of information that is uncovered during development must be uniquely assigned to one of the following:

❑ *The system under development (SUD)*, which can be actively changed by the development team and hence defines the scope of the engineering process. Note that changes in the system scope may create changes in the context of other systems if the changed property is a relevant part of the context of other systems.

❑ *The system's context*, which constrains the system development or interacts with the system during operation, as outlined above. The context is not within the scope of the engineering activities and therefore cannot be changed by the developers. However, if the context does change in some way, these changes may have an impact on the SUD [Gong 2005].

❑ *The system's irrelevant environment*, which comprises those aspects of the context that neither interact with the SUD during operation nor influence the development process. Information of this kind is irrelevant for the engineering process.

Note that changes in the scope of the SUD necessarily result in changes in the context, as other aspects of the context become relevant or irrelevant. Note also that there may be transitive influences. For example, from the perspective of the light control unit (LCU) from Chapter 2, the adaptive cruise control (ACC) is irrelevant as it does not influence the LCU. However, the ACC impacts the brake system and may thereby trigger the activation of the hazard warning light. In such cases, it might be advantageous to explicitly document those aspects. Another possibility is the existence of multiple systems of the same type within a system's context, as is the case, for example, for a collision avoidance system, which interacts with other collision avoidance systems during operation (cf. [Daun et al. 2015a]).

In order to document the relevant context aspects, the SPES XT context modeling framework [Daun et al. 2016] distinguishes between two types of context information:

Types of context information

❑ *Knowledge sources*, which provide information about the system and its operational context, such as standards, laws, regulations, etc. These knowledge sources constrain the development as outlined above and can be documented using the context of knowledge model, as illustrated in Section 4.2.1.

❑ *Entities and functions*, which interact with the system during operation. These may include human users, external systems, or non-physical services providing input to the system and receiving output from the system and which hence exchange context measurements.

The following subsections explain the specific context models of the SPES XT context modeling framework by means of illustrative examples. Afterwards, Section 4.2.3 discusses the integration of the context models into the SPES XT modeling framework.

4.2.1 The Context of Knowledge

A prerequisite for discussing context information is that knowledge sources, their relationships with one another, and their relationships to the system are properly documented. This is necessary to detect and resolve conflicts and contradictions between context information from different knowledge sources and to validate context assumptions. An ontology that structures different types of knowledge sources is given in [Daun et al. 2014a]. Fig. 4-1 shows an excerpt from a context of knowledge model for the adaptive cruise control system which is part of the automotive system cluster from Chapter 2.

Document knowledge sources

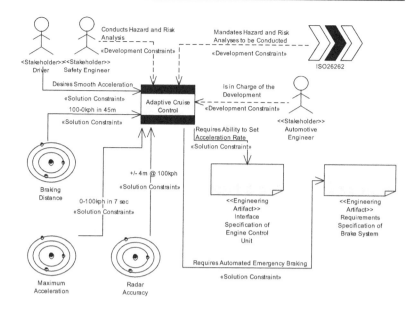

Fig. 4-1   *Context of knowledge of the adaptive cruise control (ACC) system*

Distinguish between different relationships

As we can see from Fig. 4-1, model elements can be associated with one another through either dashed arrows (denoting a constraint on the development process of the SUD) or solid arrows (denoting a constraint on the solution space of the SUD). The context of knowledge model differentiates the following types of model elements:

❑ *The system under development*, which is depicted by a rectangular shape featuring black side bars in the center of the diagram. Incoming associations depict specific information which constrains the SUD's development.

❑ *Stakeholders*, such as development team members, customers, users, and other people involved who provide information about the system's intended use, limiting factors regarding the development process, as well as the solution space. Stakeholders are represented by stick figures and denote the specific role through stereotypes.

❑ *Engineering artifacts* of external systems in the operational context or predecessor systems of the SUD (see Section 4.2.2), such as requirements, functional architectures, logical architectures, or technical architectures, test cases, manuals, or code. Because external systems interact with the SUD during operation, the development of the SUD may be constrained by, for example, interface requirements of the external systems. Engineering artifacts are depicted by rectangular shape with a turned down corner (sometimes referred to as a dog-ear) and denote the specific type through stereotypes.

❑ *Physical processes and measurements*, which document physical laws or limitations the system is constrained by. These are documented by a shape depicting an atom.

❑ *Process requirements and other documents*, which constrain the development process by demanding the execution of particular procedures or analyses or organization-specific quality gates.

Any general-purpose modeling language that provides symbols for modeling entities and their relationships, such as SysML block diagrams, can be used to create context of knowledge models without using the specific notation suggested above. However, because such modeling languages are often used to model properties of the SUD (e.g., the architecture or design), it can be difficult to distinguish between context of knowledge models and models of the SUD. Therefore, we recommend the use of dedicated modeling elements for the context of knowledge. In Fig. 4-2, we show a number of dedicated modeling elements for the context of knowledge of our example specification.

One of the core advantages of model-based development is being able to easily create views of a model (cf. [Finkelstein et al. 1992]), particularly when dealing with very complex systems, such as the automotive system cluster from Chapter 2. Because the context of knowledge model is prone to accumulating a magnitude of associations between the SUD and the knowledge sources, it may be useful to generate views which hide less relevant information while keeping the most important information for a particular purpose [Finkelstein et al. 1992]. Similarly, in addition to limiting the scope of a context of knowledge model, it might be useful to enhance its scope by including knowledge sources from the context of knowledge of other systems in the context of the SUD, as shown in Fig. 4-2.

Abstraction and landscaping of knowledge sources

As we can see, the context of knowledge of the ACC from Chapter 2 has been extended by knowledge sources from the context of knowledge of the light control unit. This creates a landscape of knowledge sources for the whole automotive system cluster, which helps to reveal transitive dependencies between knowledge sources.

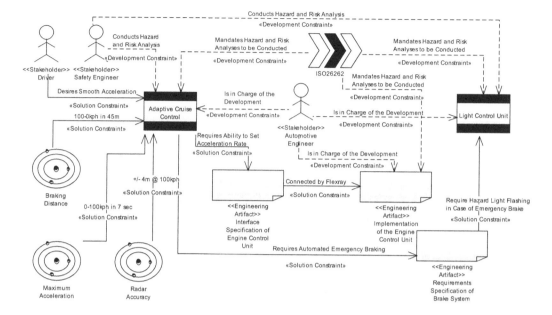

Fig. 4-2   *Landscape of the context of knowledge of the ACC*

4.2.2   The Operational Context

Interaction with the SUD during operation

The context of a system consists not only of knowledge and its sources, as illustrated in Section 4.2.1, but also of physical entities, such as users and external systems which interact with the SUD, and of these entities' functions which use the logical functions of the SUD or vice versa. Therefore, the SPES XT context modeling framework differentiates between three types of operational context models.

Structural Operational Context

Static-structural dependencies

The structural operational context focuses on physical entities that interact with the context subject, their dependencies with the context subject, as well as their dependencies among each other. For the ACC from Chapter 2, the light control unit (LCU) is a *context entity* (i.e., a context object which represents a *system in the context*). Dependencies between the ACC and the LCU include, for example, the actuation of brakes and speed measurements. However, generalization, aggregation, and composition dependencies can also be depicted.

SysML block definition diagrams can be used to document information on the structural operational context. An example from the structural operational context for the ACC is shown in Fig. 4-3, which not only depicts the dependencies of the ACC with several context entities such as the LCU or the electronic stability control, but also dependencies between context entities such as the use of the engine control unit by the electronic control unit to decelerate the vehicle.

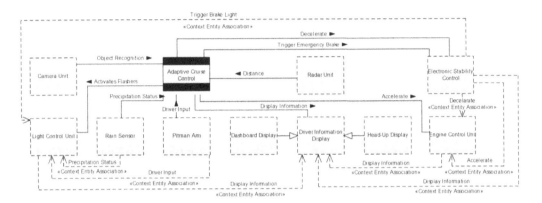

Fig. 4-3   *Structural operational context of the ACC*

Functional Operational Context

The functional operational context abstracts from physical entities and focuses on their respective services. In function-centered engineering, it is common to develop abstract conceptual functions — as realized in the functional viewpoint (see Chapter 8) — before they are divided into technical architectures (see [Daun et al. 2014b]). The functional operational context therefore allows documentation of the dependencies between the conceptual functions of the SUD and services in that the system's context which could be accessed. The functional operational context can also be used to document data flows and control flows between the SUD and *context functions*. In the automotive system cluster example from Chapter 2, the ACC function *Control (de)acceleration* uses the function *Flash hazard lights* because this function is provided by the LCU and not by the ACC. The function *Flash hazard lights* is hence considered a *context function* from the ACC's point of view. Fig. 4-4 shows this dependency and others as an SPES function network. Details of the SPES function network and more detailed examples can also be found in Chapter 9.

Conceptual functions and services

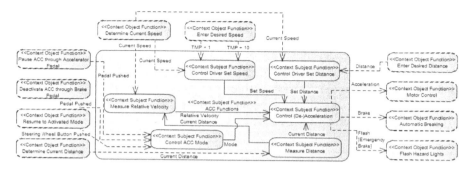

Fig. 4-4   *Functional operational context of the ACC*

Behavioral Operational Context

Identify externally visible behavioral states

In contrast to the structural or functional operational context, the behavioral operational context does not depict dependencies between the SUD and interacting context objects; instead, it shows externally observable behavior of context entities and context functions (i.e., the black-box behavior of entities and functions). Therefore, the behavioral operational context uses traditional behavioral models such as finite state automata or SysML state machines to enrich the structural or functional operational context by allowing developers to document the specific behavior of context entities or context functions respectively. Fig. 4-5 shows the externally visible behavior of the context entity *Engine Control Unit* (see Fig. 4-3).

4.2.3   Integration into the SPES Modeling Framework

Explicit context diagrams

Each SPES viewpoint across all granularity layers can use the different types of context models introduced in Sections 4.2.1 and 4.2.2. For this purpose, *explicit* context diagrams must be created as outlined above. For example, in the requirements viewpoint of the ACC, a structural operational context model could be used to depict all human users interacting with the ACC — e.g., the driver or the mechanic — as well as external systems such as the LCU. Note that the LCU is in the context of the ACC (and vice versa), while at the same time, from the perspective of the overall development process of the automotive system cluster, both systems are subsystems. For another example, the functional viewpoint could use the functional operational context model depicted in Fig. 4-4.

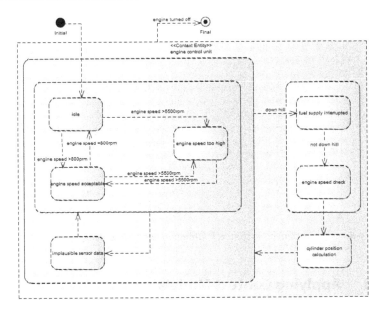

Fig. 4-5   *Behavioral operational context of the ACC*

However, we can also look for *implicit* context information in existing SPES artifacts in all viewpoints across all granularity layers. For example, scenario diagrams in the requirements viewpoint of the ACC contain the interaction between the ACC and the LCU and therefore document implicit context information. In the technical viewpoint, the interface behavior of a technical component subsumes assumptions about the externally visible behavior of some other technical component.

Implicit context diagrams

From the perspective of a particular subsystem (e.g., the ACC), context information pertaining to this subsystem (see the gray dotted rounded edge box in Fig. 4-6) may therefore be found in the viewpoints documented for other subsystems, for example, the LCU (see the grey solid rounded edge box in Fig. 4-6). Of course, relevant information, especially with regard to the interfaces, will also be found in the supersystem's viewpoints — for example, the automotive system cluster.

Context in existing SPES XT artifacts

This variety of documented context information allows several different types of analyses and quality assurance activities to support the consistent development of complex embedded systems across a multitude of degrees of granularity, as briefly outlined in Section 4.3.

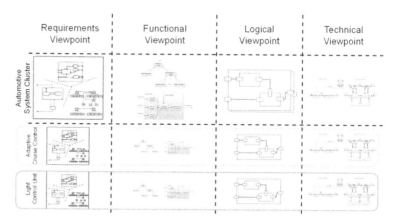

Fig. 4-6   *Documenting context information and their sources*

4.3   Applying Context Models

The SPES XT context modeling framework allows automated analysis, validation, and verification techniques to be applied in embedded systems which are developed across different granularity layers.

Change management and impact analysis

For example, documenting context information allows change impact analyses and change management. Changes in existing SPES artifacts – for example, changes within the functionality, timing behavior, or interface behavior of some subsystems – can occur throughout engineering and the current SUD may be strongly affected. In addition to changes during design time, changes can also occur during runtime. If those changes can already be predicted at design time, they should be documented as well [Daun et al. 2015b]. Based on documented context information, it is possible to trace changes relevant for the SUD (e.g., the ACC) from other subsystems' artifacts (e.g., the LCU). Conflicts resulting from these changes can be detected automatically, for example, by means of contract-based interface specifications (see Chapter 6).

Semi-automatic artifact generation

Furthermore, automated model evolution techniques can be applied to generate initial, consistent engineering artifacts of one subsystem (e.g., the ACC) from the relevant context aspects already documented within the artifacts of other subsystems (e.g., the LCU). For example, an approach to include existing contextual information into the functional design is proposed in Chapter 9.

Validation and verification

The SPES XT context modeling framework also allows a system to be validated and verified against its context. Such quality assurance often requires that development artifacts are compared against either implicit stakeholder knowledge (e.g., in the case of reviews) or against some existing specification (e.g., in the case of simulative validation or real-time verification, see Chapter 11).

Based on the context models, the current context of the system can be generated automatically from the other systems' specifications. Therefore, common techniques such as model checking can be applied to the up-to-date context model and the current system model. These techniques can also be used to support structured reuse and variability management (see Chapter 10).

Structured reuse and variability management

4.4 Summary

In this chapter, we outlined the importance of explicit documentation of context models in the engineering of embedded systems. We proposed a framework for context modeling featuring explicit documentation of different context information in four distinct but related perspectives:

❑ The *context of knowledge model* documents information sources constraining the development of the system.
❑ The *structural operational context* documents static-structural relationships to interacting context entities.
❑ The *functional operational context* documents conceptual functions of context entities accessed by the SUD and vice versa.
❑ The *behavioral operational context* allows the documentation of the externally visible behavioral properties of the structural and functional operational contexts.

We outlined how explicit and implicit context information can be uncovered in existing SPES engineering artifacts across all viewpoints and degrees of granularity.

4.5 References

[Alfaro and Henzinger 2001] L. de Alfaro, T. Henzinger: Interface Automata. In: Proc. 8th European Software Engineering Conf. and 9th ACM SIGSOFT Int. Symp. on Foundations of Software Engineering (ESEC/FSE-9), ACM, New York, 2001, pp. 109–120.

[Bergh and Coninx 2006] J. Bergh, K. Coninx: CUP 2.0: High-Level Modeling of Context-Sensitive Interactive Applications. In: Model Driven Engineering Languages and Systems, Springer, Berlin Heidelberg, 2006, pp. 140–154.

[Daun et al. 2012] M. Daun, B. Tenbergen, T. Weyer: Requirements Viewpoint. In: Model-Based Engineering of Embedded Systems: The SPES 2020 Methodology. Springer, Heidelberg/New York, 2012.

[Daun et al. 2014a] M. Daun, J. Brings, B. Tenbergen, T. Weyer: On the Model-Based Documentation of Knowledge Sources in the Engineering of Embedded Systems. In: Gemeinsamer Tagungsband der Workshops der Tagung Software Engineering 2014, Vol. 1129, CEUR-WS.org, 2014, pp. 67-76.

[Daun et al. 2014b] M. Daun, T. Weyer, K. Pohl: Validating the Functional Design of Embedded Systems against Stakeholder Intentions. In: Proc. 2nd Int. Conf. on Model-Driven Engineering and Software Development, SciTePress, 2014, pp. 333-339.

[Daun et al. 2015a] M. Daun, J. Brings, T. Bandyszak, P. Bohn, T. Weyer: Collaborating Multiple System Instances of Smart Cyber-Physical Systems: A Problem Situation, Solution Idea, and Remaining Research Challenges. In: Proc. ICSE WS – Int. WS on Software Engineering for Smart Cyber-Physical Systems (SEsCPS'15), 2015, pp. 48-51.

[Daun et al. 2015b] M. Daun, B. Tenbergen, J. Brings, T. Weyer: Documenting Assumptions about the Operational Context of Long-Living Collaborative Embedded Systems. In: Gemeinsamer Tagungsband der Workshops der Tagung Software Engineering 2015, Vol. 1337, CEUR-WS.org, 2015, pp. 115-117.

[Daun et al. 2016] M. Daun, J. Brings, T. Weyer, B. Tenbergen: Fostering Concurrent Engineering of Cyber-physical Systems – A Proposal for an Ontological Context Framework: In: 3rd International Workshop on Emerging Ideas and Trends in Engineering of Cyber-Physical Systems (EITEC), IEEE Computer Society, Los Alamitos, 2016, pp. 5-10.

[Dhaussy et al. 2009] P. Dhaussy, P. Pillain, S. Creff, A. Raji, Y. Traon, B. Baudry: Evaluating Context Descriptions and Property Definition Patterns for Software Formal Validation. In: Model-Driven Engineering Languages and Systems. Springer, 2009, pp. 438–452.

[Finkelstein et al. 1992] A. Finkelstein, J. Kramer, L. Finkelstein, M. Goedicke: Viewpoints: A Framework for Integrating Multiple Perspectives in System Development. In: Int. Journal of Software Engineering and Knowledge Engineering, Vol. 2, 1992, pp. 31-59.

[Gause 2005] D. C. Gause: Why context matters - and what can we do about it? In: IEEE Software, Vol. 22, No. 5, 2005, pp. 13–15.

[Georgia 2013] State of Georgia: Motor Vehicles and Traffic - Horns, Exhaust Systems, Mirrors, Windshields, Tires, Safety Belts, Energy Absorption Systems. Official Code of Georgia: Title 40, Ch. 8, Art. 1, Part 4, 2013.

[Gong 2005] L. Gong: Contextual modeling and applications. In: Int. Conf. on Systems, Man and Cybernetics, 2005, pp. 381-386.

[ISO/IEC 25010] ISO/IEC: ISO/IEC 25010:2011-03 – Systems and software engineering - Systems and software Quality Requirements and Evaluation (SQuaRE) - System and software quality models. 2011.

[Jackson 2006] M. Jackson: Problem frames. Addison-Wesley, Harlow, 2006.

[Jin and Liu 2006] Z. Jin, L. Liu: Towards Automatic Problem Decomposition: An Ontology-Based Approach. In: Proc. Int. WS on Advances and Applications of Problem Frames, ACM, 2006, pp. 41–48.

[Strang et al. 2003] T. Strang, C. Linnhoff-Popien, K. Frank: CoOL: A Context
    Ontology Language to Enable Contextual Interoperability. In: Distributed Appli-
    cations and Interoperable Systems, Springer, 2003, pp. 236–247.

[StVZO 2013] Federal Republic of Germany: Scheiben, Scheibenwischer, Scheiben-
    wascher, Entfrostungs- und Trocknungsanlagen für Scheiben. StVZO §40 Absatz
    1, 2013.

[Van Lamsweerde 2010] A. Van Lamsweerde: Requirements engineering. Wiley,
    Chichester, 2010.

[Yu 1995] E. Yu: Modelling Strategic Relationships for Process Reengineering. Dept.
    of Computer Science. Ph.D. Thesis, Univ. of Toronto, 1995.

[Zave and Jackson 1997] P. Zave, M. Jackson: Four Dark Corners of Requirements
    Engineering. TOSEM, Vol. 6, No. 1, 1997, pp. 1–30.

Wolfgang Böhm
Stefan Henkler
Frank Houdek
Andreas Vogelsang
Thorsten Weyer

5

SPES XT Systems Engineering Extensions

The SPES modeling framework allows seamless, model-based development of complex embedded software. It defines structures, content, and concepts used in artifacts. One major challenge when engineering embedded systems is to consider the synchronization between different engineering disciplines at the process and artifact level in a coherent manner. In this chapter, we present an extension for the SPES modeling framework supporting the close integration of systems engineering based on ISO/IEC 15288 with software engineering based on ISO/IEC 12207 and software-related engineering disciplines by using the core concepts of the SPES modeling framework as the foundation for a general engineering philosophy for the architectural design of embedded systems.

© Springer International Publishing AG 2016
K. Pohl et al. (eds.), *Advanced Model-Based Engineering of Embedded Systems*,
DOI 10.1007/978-3-319-48003-9_5

5.1   Introduction

Challenge   A multitude of disciplines are involved in the engineering of embedded systems. One major challenge here is to consider the synchronization between different engineering disciplines at the process and artifact level in a coherent manner. Process standards address this challenge by defining a transition between the activities involved from the disciplines but they do not provide sufficient support with respect to the corresponding artifacts and their relationships. The missing artifact-oriented integration of the engineering disciplines (e.g., electrical engineering, software engineering, and mechanical engineering) within the systems engineering process leads to error-prone and cost-intensive synchronizations.

Integration of engineering disciplines   Existing standards which aim to integrate engineering disciplines at the systems and software engineering process level — such as ISO/IEC 12207 [ISO/IEC 12207], ISO/IEC 15288 [ISO/IEC 15288], and ISO/IEC 29148 [ISO/IEC/IEEE 29148] — typically try to achieve integration by proposing activities and corresponding relationships between them within the related disciplines. This means that for each activity, we define which other activities need to be performed as a prerequisite because these activities provide necessary input (see also Section 3.3). Typically, existing process approaches that also consider artifacts for describing the input and output of the activities do this only at an abstract level (e.g., V-Model_XT [Rausch et al. 2005]). On the other hand, artifact-oriented models that are defined in a formal way (e.g., [Gausemeier et al. 2007, Braun et al. 2010]) lack support for a recursive system structure as defined in ISO/IEC 15288. Furthermore, these approaches typically focus only on relationships between some specific artifact types from different disciplines.

Approach for integration   In this chapter, we present an approach that enables close integration between the different engineering disciplines based on standard processes using the core concepts of the SPES XT modeling framework. The underlying artifact model of the SPES XT modeling framework facilitates the seamless integration of the corresponding systems and engineering processes. The explicit concept of viewpoints and granularity layers combined with the artifact model enables the SPES XT modeling framework to also support the required recursive engineering of embedded systems.

Usually, systems engineering considers the complete life cycle of a system, including, for example, the operating phase of the system. In this chapter, we focus on the architectural design phase within the life cycle. The approach presented extends the work of Böhm et al. [Böhm et al. 2014] by considering the integration of the different disciplines in a more general way.

The chapter is structured as follows: Section 5.2 introduces the basic principles for the approach and Section 5.3 describes the mapping between systems and software engineering activities by applying the SPES XT modeling framework.

5.2 Standard Engineering Processes

ISO/IEC 15288 takes a very broad view and can be applicable for the engineering process of any technical system. To manage the size and complexity of a system during the engineering process, the system is divided into subsystems that are designed independently and finally integrated into the overall system. A subsystem itself can again be considered as a system. Therefore, the recursive application of ISO/IEC 15288 leads to a hierarchy of systems, subsystems, and finally basic system elements. As a result, the SUD is decomposed into a structure of manageable and understandable system elements which can be implemented, reused, or acquired from another party. Note that the number of hierarchy levels depends on the system being developed. However, not all subsystems in the hierarchy need to have the same number of hierarchy levels.

ISO/IEC 15288

For the implementation of the system elements (i.e., realization of the implementation subprocess of ISO/IEC 15288), domain-specific processes may be applied. In the engineering of embedded systems, some system elements consist of software only. The engineering of a technical platform and the integration of software items into this hardware environment are addressed in ISO/IEC 15288. For example, software items can be implemented using the framework for software life cycle processes as defined in ISO/IEC 12207.

In this context, software is treated as an integral part of the system which performs certain functions within the system. The software is developed by identifying and refining the software requirements from the system requirements, implementing the software, and integrating it into the overall system design. It is important to note that the order of these processes is not necessarily sequential.

Software as an integral part

The output of the processes on a higher system hierarchy level is the input for the processes of the level below.

5.3   Integrating Systems and Software Engineering

When developing the software for an embedded system, it is not only the integration of the software engineering view into the general systems engineering view that needs to be considered; the engineers also have to take into account the relationship between the software items and items of other disciplines like electrical and mechanical engineering.

Two orthogonal dimensions   As described in Chapter 3, the SPES XT modeling framework supports the development of software for embedded systems by differentiating between the following two orthogonal dimensions: viewpoints and granularity layers. Based on the SPES XT context modeling framework described in Chapter 4, in this chapter we describe how to integrate standard systems engineering processes with the discipline-specific engineering processes.

5.3.1   Mapping of the SPES XT Modeling Framework to Standard Engineering Processes

When mapping the SPES XT modeling framework to ISO/IEC 15288 and ISO/IEC 12207, we have to consider two dimensions: first, we need to map the different subprocesses defined in the standards to artifact types of the SPES XT modeling framework. We then have to understand how the SPES XT modeling framework can support the characteristics of the standards. Each process defined in the two standards defines a set of activities and lists their outcomes.

The artifacts defined in the SPES XT modeling framework can be used as a reference capturing the results of the processes. Just like in artifact-oriented development, the content items are independent of the development process and the artifact model is mapped to the development process and vice versa [Böhm and Vogelsang 2013]. This mapping is achieved by filtering the results of the process activities and abstracting from the methods to produce the results. This allows us to assign the activities and milestones of the process to artifact types of the SPES XT modeling framework (see Chapter 3). The main reason this assignment works is due to the concept model and the strict separation of content and structure of the artifacts, which makes the SPES XT modeling framework independent of the development processes applied. In the following table, we present a high-level mapping of the process steps defined in ISO/IEC 15288 and ISO/IEC 12207 to the artifact types of the SPES XT modeling framework.

Tab. 5-1　　　*Mapping process standards and SPES XT artifact types*

Standard	Process step	Artifact types of the SPES XT modeling framework
ISO/IEC 15288 (systems engineering)	Stakeholder requirements definition	Context model (requirements viewpoint), goal model (requirements viewpoint), scenario model (requirements viewpoint)
	System requirements analysis	Solution-oriented requirements (requirements viewpoint), functional black box model (functional viewpoint)
	System architecture design	Functional white box model (functional viewpoint), logical architecture (logical viewpoint), technical architecture (technical viewpoint)
ISO/IEC 12207 (software engineering)	Software requirements analysis	Context model (requirements viewpoint), goal model (requirements viewpoint), scenario model (requirements viewpoint), solution-oriented requirements (requirements viewpoint), functional black box model (functional viewpoint)
	Software architecture design	Functional white box model (functional viewpoint), logical architecture (logical viewpoint)
	Software detailed design	Logical architecture (logical viewpoint), technical architecture (technical viewpoint)

As shown in the table, the process defined in ISO/IEC 15288 comprises requirements and design phases for software development which are introduced by ISO/IEC 12207. This suggests progressing in a strictly sequential manner. However, in practice, this sequential course is suspended by interleaving the requirements and design

ISO/IEC 15288 comprises requirements and design phases

processes of ISO/IEC 15288 and ISO/IEC 12207, which is a violation of the strict process orientation of the standards.

5.3.2   Integrating Engineering Processes Using the SPES XT Modeling Framework

As introduced in Section 5.3.1, we apply the two dimensions of the SPES XT modeling framework together with the artifact types associated with the process steps of the standards and the context modeling approach.

Solution concept   Fig. 5-1 illustrates our solution concept. The upper part illustrates the systems engineering activities mapped to the SPES XT modeling framework. The system model is developed on different system granularity layers and different viewpoints until single system elements fulfill the requirements for a mapping to a specific engineering discipline. Typically, this system-level solution is part of the technical viewpoint because the discipline-specific properties characterize this model with discipline-specific components. Based on this system-level solution, the different engineering disciplines start the discipline-specific development in parallel. The system-level solution also represents the relationship between the different components of the different disciplines.

The discipline-specific components characterized are the system under development for the different discipline-specific engineering paths, which start again in the requirements viewpoint. All elements of other disciplines are part of the context of the SUD. Furthermore, discipline-specific context elements, such as the user or sensors, are also modeled.

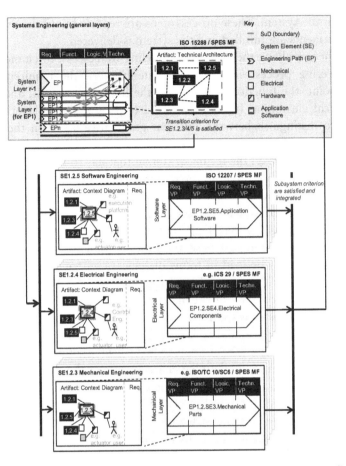

Fig. 5-1   *Integration of ISO/IEC 15288-compliant and ISO/IEC 12207-compliant engineering processes using the SPES XT modeling framework*

The discipline-specific engineering can be guided by a standard process (e.g., ISO 12207 for software engineering, ICS 29 for electrical engineering, and ISO/TC 10/SC6 for mechanical engineering). The SPES XT modeling framework approach enables an early detection of violations between the different engineering disciplines using the artifacts and their relationships which span the disciplines. If an interface or a property to another discipline is violated, the discipline-specific solution or, in the worst case, the system-level solution, must be changed. Note that this could be a labor-intensive and cost-intensive process. Therefore, it is important to detect such violations at an early stage. If all discipline-specific requirements are fulfilled and the discipline-specific solutions are integrated, an inte-

SPES XT modeling framework enables early detection of violations

grated system-level solution is developed — for example, a prototype of the system. These process steps are repeated until the overall
requirements are fulfilled and the system is delivered.

Application of the SPES
XT modeling framework

When applying the SPES XT modeling framework to (standard)
engineering processes, we differentiate between the requirements
viewpoint, the functional viewpoint, the logical viewpoint, and the
technical viewpoint. In addition, we use the granularity layer concept to distinguish between different granularity layers in the systems engineering process.

First layer

In the first layer, the technical system that has to be developed is
regarded as a whole. Within the requirements view, both the operational context and the requirements for the system are documented.
In the functional view, the user functions and the corresponding
system functions are specified from a structural and a behavioral
perspective. In the logical view, the structure of a specific conceptual architecture is defined in terms of logical components and their
relationships. Within the technical view, the specific technical solution — which consists of technical components, their properties,
and relationships — is specified.

Subjacent granularity
layer

Following the granularity layer concept of the SPES XT modeling framework, the transition to a subjacent granularity layer is
defined by the structure-defining characteristic of the corresponding
granularity layer. Typically, the logical architecture (part of the
logical viewpoint) or the technical architecture (part of the technical
viewpoint) is used as the structure-defining characteristic for the
corresponding engineering paths in the subjacent granularity layer
(see Section 3.2.2). The structure-defining characteristic defines a
set of logical or technical components realized within dedicated
engineering paths in the subjacent granularity layer of the system.
Note that the source of the structure-defining characteristic can
change between different granularity layers and even between different engineering paths within one granularity layer.

Example: hazard
warning light

In our example, the technical architecture of the hazard warning
light (a subsystem of the exterior lighting system, see Section 2.2.2)
describes which technical components (e.g., electronic control units,
ECUs) are involved in implementing this feature. Fig. 5-2 shows the
subset of the technical architecture of the vehicle (also called E/E
architecture, see also Section 2.2.3) along with the required components. Each technical component of the hazard warning light system, such as the upper control panel (UCP), is considered in a dedicated engineering path in the subjacent granularity layer.

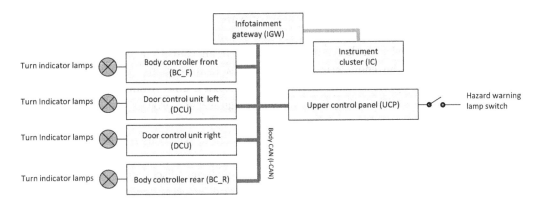

Fig. 5-2   *Technical architecture of the hazard warning light system*

In ISO/IEC 15288, the physical architecture in the lowest granularity layer consists of system elements that are developed by a set of activities defined by the standard without decomposing the corresponding system element into detailed parts that are engineered within separate engineering paths in a subjacent granularity layer. When applying the SPES XT modeling framework to ISO/IEC 15288-compliant systems engineering processes, system elements are resident either in the logical architecture within the logical view or the technical architecture within the technical view. Fig. 5-3 shows the logical architecture of the hazard warning light system in the lowest granularity layer of the ISO/IEC 15288-compliant systems engineering process and the corresponding system elements in terms of logical components.

Discipline-specific processes based on the SPES XT modeling framework

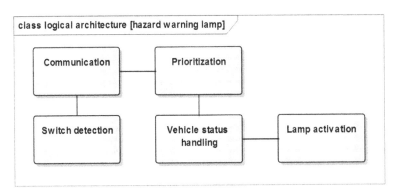

Fig. 5-3   *Logical architecture of the hazard warning light system*

*Realization in an
interaction of different
systems engineering
disciplines*
Typically, a system element consists of hardware (e.g., mechanical or electrical parts) and software and has to be realized via an inter-action of different systems engineering disciplines (e.g., mechanical, electrical, control, software). If all transition criteria are satisfied, the engineering focus changes from systems engineering to disci-pline-specific engineering processes; for instance, to an ISO/IEC 12207-compliant software engineering process for developing the application software of the corresponding system element.

In the remainder of this section, we focus on the transition be-tween the general systems engineering process, the software engi-neering, and the electrical engineering discipline. Nevertheless, the software and electrical engineering process depend on the results of the other disciplines (e.g., mechanical engineering or control engi-neering) and vice versa. This means that during the detailed engi-neering of a system element, a close cooperation between the differ-ent engineering disciplines is required which has to be coordinated by the overall systems engineering discipline. These dependencies are modeled in the context of the software element in the SPES XT modeling framework.

*Starting point of the
discipline-specific en-
gineering processes*
The starting point of the engineering processes for a system ele-ment is the logical or technical architecture of the engineering path in which the transition decision for the corresponding system ele-ment has been made (see Fig. 5-1). The technical architecture doc-uments the relationship of the system element in question to other system elements. In addition, the context diagram of the require-ments view in that engineering path provides important information about the relationship of the focused system element to system ele-ments on a high-level granularity layer and the operational envi-ronment of the overall technical system. Fig. 5-4 shows the context diagram of the application software for handling the vehicle status information to determine the light-dark ratio duration of a blinking cycle.

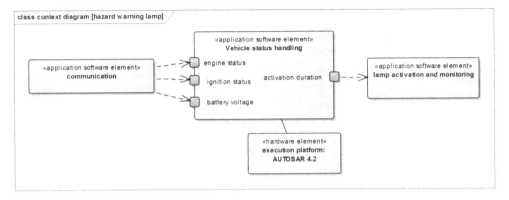

Fig. 5-4   *Context diagram of the application software element "Vehicle status handling"*

In the software engineering process, we consider the systematic development of the application software for the corresponding system element. The application software runs on an execution platform which consists of hardware (e.g., ECUs, CPU, memory) as well as platform-related software (e.g., operating system software, device drivers).

Systematic development of application software

Following the philosophy of the SPES XT modeling framework, in the software engineering discipline, the context diagram of the corresponding SUD (i.e., the application software) that documents the environment in which the application software will operate when the overall embedded system is in operation has to be created (see Fig. 5-4). The context diagram shows all the operational relationships of the application software to other systems, other system elements, and the operational constraints that are defined by other engineering disciplines. Based on this context diagram, the goals and corresponding scenarios of the application software are elicited, analyzed, and documented in the next step. The specification of the detailed requirements for the application software is based on the context diagram as well as on the goals and scenarios of the requirements view.

Context diagram documents operational environment

The application of the SPES XT modeling framework in the other engineering disciplines, such as electrical and mechanical engineering, follows the same principles. Of course, at the visual syntax level, the discipline-specific diagrams differ a lot. While we typically use class diagrams (or component diagrams) and, for example, state machines for modeling the structure and behavior of the logical and technical viewpoint of a software application, wiring diagrams are

Application of the SPES XT modeling framework in other engineering disciplines

preferred for the electrical engineering discipline. Fig. 5-4 shows the actual hazard warning light switch with its embedded indicator lamp. It shows the components of the circuit as simplified shapes and the power and signal connections between the devices. The behavior is specified by the logical operations of the switches. The relationship between the two disciplines shown is given based on the system-level technical architecture (see Fig. 5-2) by the hazard warning light switch element and by the control units.

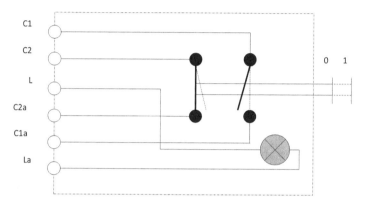

Fig. 5-5   *Wiring diagram of the hazard warning light switch*

5.4   Summary

In this chapter, we presented the SPES XT modeling framework as a model-based approach that bridges the gap between the system and software engineering processes of embedded systems. We related basic activities of standard engineering processes to artifact types defined in the SPES XT modeling framework. In this way, we contribute a solution to the problem that the engineering disciplines may have different ideas on how to structure the underlying input and output artifacts. Using the SPES XT Systems Engineering Extensions, we increase consistency between the system and the software engineering artifacts and enable capabilities for tracing changes and performing automated analyses and transformations, which in turn lead to increased effectiveness and efficiency.

5.5   References

[Böhm and Vogelsang 2013] W. Böhm, A. Vogelsang: An Artifact-Oriented Framework for the Seamless Development of Embedded Systems. In: Software Engineering (Workshops), 2013, pp. 225 - 234.

[Böhm et al. 2014] W. Böhm, S. Henkler, F. Houdek, A. Vogelsang, T. Weyer: Bridging the Gap between Systems and Software Engineering by Using the SPES Modeling Framework as a General Systems Engineering Philosophy. In: Elsevier Procedia Computer Science, Vol. 28, 2014, pp. 187-194.

[Braun et al. 2010] P. Braun, M. Broy, F. Houdek, M. Kirchmayr, M. Müller, B.
Penzenstadler, K. Pohl, T. Weyer: Guiding requirements engineering for software-intensive embedded systems in the automotive industry. In: Computer Science - Research and Development, Vol. 29, No. 1, 2010, pp. 21-43.

[Gausemeier et al. 2007] J. Gausemeier, H. Giese, W. Schäfer, V. Axenath, U. Frank,
S. Henkler, S. Pook, M. Tichy: Towards the Design of Self-Optimizing Mecha-
tronic Systems: Consistency between Domain-Spanning and Domain-Specific
Models. In: International Conference on Engineering Design, ICED'07, August,
Paris, France, 2007, pp. 28-31.

[ISO/IEC 12207] ISO/IEC: ISO/IEC 12207:2008 – Systems and software engineering
-- Software life cycle processes. Edition 2, 2008.

[ISO/IEC 15288] ISO/IEC: ISO/IEC 15288:2008 – Systems and software engineering
-- System life cycle processes. Edition 2, 2008.

[ISO/IEC/IEEE 29148] ISO/IEC/IEEE: ISO/IEC/IEEE 29148:2011 – Systems and
software engineering -- Life cycle processes -- Requirements engineering. 2011.

[Rausch et al. 2005] A. Rausch, C. Bartelt, T. Ternité, M. Kuhrmann: The V-Modell
XT Applied–Model-Driven and Document-Centric Development. In: 3rd World
Congress for Software Quality, Vol. 3, 2005, pp. 131-138.

Part III

Application of the SPES XT Modeling Framework

Karsten Albers
Peter Battram
Alfred Bognar
Veronika Brandstetter
Andreas Froese
Bastian Tenbergen
Andreas Vogelsang
Joachim Wegener

6

Early Validation of Engineering Artifacts

Valid engineering artifacts are a key success factor for reliable (software) systems. Validity in this sense means that the engineering artifacts are the right ones to completely and correctly describe system properties as desired by stakeholders, and that they are consistent among one another. It is widely acknowledged that the later the engineering artifacts and results are validated, the higher the effort for correcting any defects found will be, leading to serious budget overruns and project delays. In this chapter, we identify major challenges of validation in early phases of the development process, including reducing the ambiguity of natural language, handling complex systems, and the need for a more structured and automated approach to validation. We introduce the requirements quality assessment framework (RQAF) as a structured approach for performing validation systematically. Furthermore, we describe a set of methodological building blocks that instantiate the RQAF and provide exemplary solutions for the stated challenges.

© Springer International Publishing AG 2016
K. Pohl et al. (eds.), *Advanced Model-Based Engineering of Embedded Systems*,
DOI 10.1007/978-3-319-48003-9_6

6.1   Introduction

Validation of engineering artifacts to ensure quality

The purpose of validating engineering artifacts is to ensure that the engineering artifacts are of sufficient quality for subsequent development activities. In particular, artifacts must reflect stakeholder intentions completely and correctly with regard to the system under development (SUD) and must be consistent among each other. Hence, these artifacts can be considered as the (requirements) specification and are subject to validation, that is, ensuring that a system is adequate with regard to some operational purpose [Boehm 1981]. There is a plethora of additional quality attributes [ISO/IEC 25010] that artifacts should satisfy. If engineering artifacts do not satisfy these quality criteria, the consequence may be serious project delays or budget overruns [Boehm 1981]. Therefore, it is essential to validate all engineering artifacts at the earliest possible stage during development.

Many forms of validation

Validation can take many forms. For example, requirements can be validated with regard to the stakeholder's vision of the finished product, to assess whether the right requirements have been elicited, or to assess whether safety or security measures are adequate [Tenbergen et al. 2015]. Furthermore, intended or unintended interactions between the system and its context (e.g., human users or other systems) can be validated to ensure that the interaction leads to the right outcome, or design decisions documented in logical or technical component models can be validated to see whether they satisfy specific qualities (e.g., performance or security).

A collection of methods

This chapter presents a collection of methods which focus on the static and dynamic validation of engineering artifacts at different granularity layers as early as possible in the system development process. More specifically, the methods introduced provide guidance for creating artifacts that can be validated right from their creation. Depending on the specific validation objective, validation faces different challenges. These challenges are briefly discussed in the next section. To meet these challenges, this chapter presents a generic validation framework (Section 6.1.2) that structures validation activities and associates artifacts necessary to validate engineering artifacts of the SPES XT modeling framework (Section 6.2). This is followed by the presentation of a selection of methodological building blocks (see Section 6.3) that produce and/or consume requirements and modeling artifacts to support validation in early phases of development.

6.1.1 Challenges for Early Validation

Validation has often been described as one of the most critical, yet most daunting tasks during system development [Boehm 1981]. The main vehicles for validation are manual techniques, such as Fagan inspections, desk-checks, or group reviews [Wiegers 2003]. However, manual validation involves a high degree of subjectivity. Traditionally, the outcome of validation depends largely on the understanding of the stakeholders involved regarding the problem domain [Glinz and Fricker 2014], [Gacitua et al. 2009] and their understanding of the development process [Lisagor et al. 2010], [Shull et al. 2000]. Furthermore, validation often lacks a clear objective and the sheer volume of artifacts to be reviewed, mostly documented using unstructured natural-language requirements [Sikora et al. 2012], leads to poor validation coverage, decreased validation effectivity and efficiency, and validation results that are hard to confirm [Flynn and Warhurst 1994]. To alleviate these issues, validation activities must be structured appropriately and ideally supported by automatic or semi-automatic techniques. Consequently, the following major challenges arise:

❑ **Increase objectivity:** Many engineering artifacts, especially in early phases, rely on textual descriptions that are subject to interpretation. Validation activities are much more precise if the ambiguity of the artifacts is kept to a minimum.

❑ **Scale validation activities:** Modern systems are specified by a large number of regularly changing artifacts to be considered during validation. (Semi-)Automatic and tool-based techniques are needed to apply validation in industrial settings with high turnover rates and frequent updates to engineering artifacts.

❑ **Handle complexity:** Modern systems are becoming increasingly complex. To this end, model-based development is often used to meet the challenges due to increased complexity. However, validation approaches must also be adapted to handle increased system complexity and the complexity in the models describing the systems.

❑ **Integrate validation:** Embedded systems are composites of mechanics, electronics, and software. Therefore, validation methods must consider the whole mechatronic system and not only an isolated part of it.

6.1.2   The Requirements Quality Assessment Framework

RQAF structures a generic validation process

To address these challenges, the *requirements quality assessment framework* (RQAF) has been developed. The RQAF (shown in Fig. 6-1) structures a generic validation process independently of a specific validation technique. In this sense, the RQAF can be used for simple reviews or inspections as well as for more sophisticated validation strategies such as automated multi-aspect validation (see Section 6.3.2) or simulative validation (Section 6.3.3).

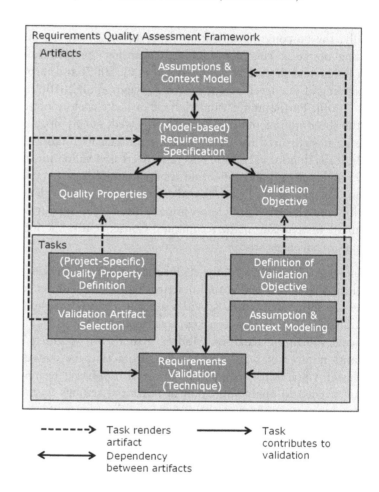

Fig. 6-1   *Structure of the Requirements Quality Assessment Framework*

The RQAF consists of a set of artifacts necessary to conduct validation and a set of validation tasks that produce these artifacts:

*RQAF consists of a set of
artifacts*

- ❑ **Validation objective:** The validation objective determines the
  goal for the validation, (i.e., why validation is being performed).
  Depending on the validation objective, developers can determine which artifacts must be subject to validation. For example,
  when validating real-time behavior, executable models could be
  used (e.g., statecharts annotated with timing constraints). The
  validation objective can be defined freely with regard to the specific needs of a development project and can be documented in
  prose text.
- ❑ **Quality property:** Quality properties are defined with regard to
  the validation objective. Therefore, quality properties cannot be
  defined in general terms but rather depend on the specific circumstances in the development project concerned. For example,
  if the validation objective is to assess the completeness of a safety argument (see Chapter 6), the quality property *completeness*
  according to [ISO/IEC 25010] could be defined as *All identified
  hazards are mitigated by at least one safety goal, which in turn
  has been refined by at least one functional safety requirement.*
  Like validation objectives, we define quality properties in prose
  text.
- ❑ **Assumptions and context models:** In contrast to many other
  quality properties, the validity of engineering artifacts depends
  largely on stakeholders' intentions and understanding of the
  problem domain (see Section 6.1.1 as well as [Tenbergen et al.
  2015]). This information is typically documented poorly or not
  at all. However, the validity of engineering artifacts can only be
  assessed based on some type of reference artifact. The SPES XT
  context modeling framework (see Chapter 4) provides means
  for modeling assumptions about the organizational or operational context, documenting assumed context entities in interaction with the system, and documenting knowledge sources that
  possess information about conditions under which artifacts are
  valid. Context models can also be used to derive specification
  artifacts and simulation configurations systematically (see Section 6.3.3 as well as [Daun et al. 2015]).
- ❑ **Requirements specification:** The requirements specification
  consists of all requirements artifacts that are subject to validation given a validation objective. This may be a subset of all
  available requirements artifacts (e.g., only validate safety-
  relevant requirements) or may require additional information to

be annotated (e.g., timing information, if not already present). For example, to validate user input, goal models are less relevant than scenario models. Requirements artifacts can be documented either in a model-based fashion (see Chapter 3) or by means of semi-formal requirements patterns (see Section 6.3.1).

❑ **Validation technique:** The validation technique is a placeholder for the specific validation method. Examples are Fagan inspections

❑ [Fagan 1976], reviews [Wiegers 2003], or any of the methods suggested in Section 6.3. Regardless of the specific technique, validation will consider the requirements specification and validate its quality against the assumptions and context models under consideration of the quality properties with regard to the specific validation objective.

Validation results are documented by artifacts

The validation results are documented by artifacts (e.g., inspection reports, analysis results, or counterexamples). Therefore, validation artifacts are defined for each individual validation process with respect to a given validation objective. Depending on the objective, quality criteria can be defined for assessing the quality of a specification. The RQAF guides the user in defining and documenting the respective validation artifacts, specifies the relationships between them, and hence supports reproducible and comprehensible results.

6.2   Supporting Artifacts for Validation

Validation techniques for a number of modeling artifacts

In this chapter, we introduce validation techniques for a number of modeling artifacts of the SPES XT modeling framework which serve as (requirements) specifications. Furthermore, we define two additional artifact types necessary for the methodological building blocks which instantiate the RQAF validation tasks. In the following, we describe *domain knowledge models* and *boilerplate specifications* as supporting artifacts and later introduce the methodological building blocks in Section 6.3. While domain knowledge models can generally be applied for the entire development project and hence are not specific to any viewpoint in particular, boilerplate specifications can be used for any type of architecture model in the functional, logical, or technical viewpoint.

6.2.1   Domain Knowledge Model

In contrast to context models (see Chapter 4), which document the assumed operational context and context of knowledge of a system, domain knowledge models comprise the available knowledge of the problem domain, similar to a project glossary (see [Daun et al. 2014] for more details on the context of knowledge). They hence make up part of the context of knowledge of the system and can be used to mitigate misinterpretation, to reduce ambiguity, and to provide a possibility for early validation of requirements specifications. Specifically, the domain knowledge is the representation of the understanding of entities and terms used by the writer of the requirements specification and/or the architect of the system model, thereby supplementing the information recorded in the viewpoint models of the SPES XT modeling framework. Thus, the domain knowledge model provides domain-specific meanings or definitions of the entities and terms used in the requirements specification and their relationships. When a new entity or term is defined in an artifact of any viewpoint, a corresponding definition of the meaning and its relationship must be created in the domain knowledge model. We describe a domain knowledge model using entities, which have a name, and labeled relationships between entities. Fig. 6-3 shows an example of a domain knowledge model. To enable an early validation of a requirements specification, the corresponding entities in the domain knowledge must also be validated (see Section 6.3.1). The requirements can only be validated if the meanings and relationships of the terms and entities in the requirements specification are correctly understood.

Domain knowledge models comprise the available knowledge of the problem domain

6.2.2   Boilerplate Specifications

Boilerplates provide a semi-formal requirements representation that can still be read like natural language but introduces a formalism which enables automatic analyses and consistency checks early in the development process, where information is typically provided in natural language. Boilerplates are a predefined set of patterns with fixed syntax. An abstract example for a boilerplate describing a system which shall perform an action after an event has occurred is as follows:

Boilerplates provide a semi-formal requirements representation

If <event>, <system> shall <action>.

Boilerplates support structured and precise capturing of requirements

Boilerplates support the requirements engineer in capturing requirements in a more structured and precise way using fixed formalization terms on the one hand, and established wording that is part of the domain knowledge (see Section 6.2.1) on the other hand. Boilerplates can hence be used to augment requirements models. Considering the ACC subsystem of the automotive system cluster example, a requirement that demands that the system is deactivated if the brake pedal is pressed is specified by a boilerplate as follows:

> If the driver presses the brake pedal,
> the ACC shall be deactivated.

Boilerplate requirements possess a precise sentence structure that is consistent with the underlying *domain knowledge*.

6.3   Validation Techniques

To validate artifacts of the SPES XT modeling framework (see Chapter 3) and its extensions described in Section 6.2, we instantiate the RQAF (see Section 6.1.2) to build systematic validation activities which we describe with methodological building blocks. Fig. 6-2 shows how constituents of the RQAF are instantiated by methodological building blocks and references the sections in which we describe these.

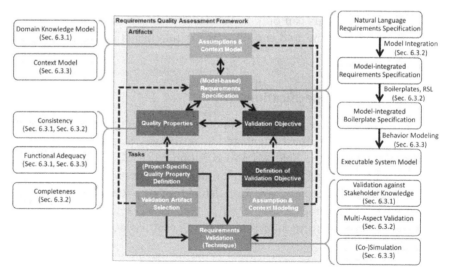

Fig. 6-2   *The RQAF is instantiated by the methodological building blocks described in the subsections of this section*

Depending on the degree of formalization, different validation techniques can be applied to validate the development artifacts. Formalization of the specifications is a prerequisite for automating these techniques to ensure compliance with quality properties such as consistency, freedom from contradictions, and completeness. Purely textual specifications can be tested thoroughly only by manual review techniques. Therefore, as part of the SPES methodology, we develop different approaches to formalize textual specifications that may be applied individually or in combination. As a technique, model integration (see Section 6.3.2) tries to combine the advantages of model-based and textual specifications and to allow validation options with regard to the aforementioned quality characteristics. Boilerplates pursue the idea of describing requirements in standardized text templates and using domain ontologies to provide testing possibilities. A much higher degree of formalization is achieved by using executable specifications (see Section 6.3.3). Here, the system behavior is described in detail using modeling languages which allow a simulation of the system behavior. For quality ("non-functional) requirements, in particular the temporal behavior of systems, we describe methods for modeling and analyzing the temporal behavior resulting from a system specification.

Degree of formalization influences the applicability of validation techniques

6.3.1 Validation against Stakeholder Knowledge

Using natural language is the most common way to document engineering artifacts [Sikora et al. 2012]. However, as outlined in Section 6.1.1, the validation of natural language artifacts is burdened by the sheer volume of requirements as well as poor structuring and subjective assessments. The following method (Tab. 6-1) is a *validation technique* with the *objective* of validating the feasibility, correctness, and consistency of requirements within a *requirements specification* by utilizing an explicit domain knowledge model. Thus, this method is an instantiation of the RQAF.

Natural language for documentation of engineering artifacts

Tab. 6-1   *Validation against documented domain knowledge*

SPES XT methodological building block	
ID	6.1
Name	Validation against stakeholder knowledge
Method	1) Build a domain model by identifying objects and relationships in the customer requirements specification.
	2) Compare and integrate the domain knowledge model in the domain model of the company.
	3) Identify and resolve ambiguous terms.
Input	Domain knowledge, natural language specification
Output	Analysis results
Condition/limitation	-

Validation against documented domain knowledge

For this method, we assume an existing domain knowledge model (e.g., of a supplier) that we use to validate a requirements specification against (e.g., provided by a customer). Fig. 6-3 illustrates an excerpt of such a domain model for the ACC subsystem of the automotive system cluster (see Chapter 2). The requirement specification may specify that the speed limiter of the cruise control system shall be deactivated temporarily if the gas pedal is pressed more than 90% by the driver. In our example, the interpretation of temporal deactivation of the speed limiter differs between the domain knowledge of the customer and the domain knowledge of the supplier. From the customer's point of view, the limiter entity is deactivated while the pedal is pressed more than 90%. From the supplier's point of view, the temporal deactivation of the limiter entity is deactivated until the driver releases the pedal completely.

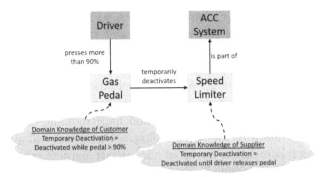

Fig. 6-3   *Excerpt of the domain knowledge model from the ACC subsystem of the automotive system cluster*

The first step of the method is to identify and collect all subjects   *Method steps*
and objects (nouns in a text) that appear in the customer require-
ments specification (e.g., *driver, gas pedal, speed limiter,* and *cruise
control system*). Then, any additional qualifying attributes or even
relative clauses are considered (e.g., *Pressed more than 90%* or
Temporarily deactivates). The collection is analyzed to identify
redundancies and synonyms. Similar or equivalent terms are
grouped under one common term. Then, the resulting terms are
classified by level of granularity and meaning and the relationships
between the terms are stated. The resulting terms are called customer entities (see Fig. 6-3 for the resulting entity relationships of the
aforementioned customer requirement). The customer entities are
compared to the domain knowledge entities from the supplier's
domain knowledge model to find the relevant entities of the supplier's knowledge. If a domain entity resembles a customer entity
(complete match), the domain entity is inserted into the domain
knowledge model. Then, the relationships between the entities are
matched to handle customer entities with no complete match. If a
customer entity has no direct match with a domain entity but has
relationships to other matching domain entities, that customer entity is a partial match and is added to the project knowledge as a new
project entity. If the customer entity neither resembles a domain
entity nor can any relationship to a domain entity be assigned (no
match), the customer entity is inserted into the project knowledge as
a new project entity using a best guess for the embodiment into the
relationship network. As a result of this process, a version of the
supplier requirements specification (SRS) can be prepared which
traces the requirements to the entities of the project knowledge and
to its corresponding customer requirement. Since the customer re-

quirements trace to the customer entities, a logical path can be established between the project entities and the customer entities. Based on this tracing, the entities of the project knowledge can be validated to clarify that the selected project entities are correct, feasible, and unambiguous. After the application of the method, the different concepts of temporal deactivation of the speed limiter are explicit (see Fig. 6-3) and the common understanding of the temporal deactivation of the speed limiter is elaborated and added to the project knowledge. Some of the project entities can contain extensions or even innovations that come from the current project. Those entities should go through an abstraction process in which any project-specific definitions are removed.

6.3.2   Artifact Structuring and Formalization to Enable Automated Analysis

A validation technique   The following method is a *validation technique,* the *objective* of which is to validate the correctness and consistency of requirements within a *requirements specification* by using boilerplate specifications that are integrated in a model that structures the specification. Thus, this method is an instantiation of the RQAF. The method is structured in three methodological building blocks: model integration (Tab. 6-2), formalization of natural language requirements (Tab. 6-3), and multi-aspect validation (Tab. 6-5). Fig. 6-4 shows the building blocks as a process of construction (solid arrows) and validation steps (dashed arrow).

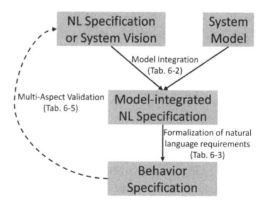

Fig. 6-4   *Overview over the validation technique*

Model integration as a technique may also be used without formalization of natural language requirements. This already increases the consistency of specifications and thus allows a lightweight validation process. Whether natural language requirements are subsequently formalized depends on the desired or necessary degree of formalization. Model integration allows the creation of semi-formal specifications, even if only sparse and inconsistent knowledge of the SUD is given. The method can be applied to informal textual specifications and produces semi-formal, model-integrated specifications that can still be read intuitively and understood by system experts not trained in the model integration approach. Model integration improves the quality of specifications and accelerates test case specification. Furthermore, it allows the definition and application of model-based completeness criteria to functional specifications, which leads to improved understandability and testability during subsequent development and validation activities.

Model integration as a technique

Tab. 6-2   *Model integration*

SPES XT methodological building block	
ID	6.2
Name	Model integration
Method	1) Create a specification chapter for each element of the model. 2) Supplement the specification with conditional expressions for each chapter.
Input	Natural language specification or vision, structural or behavioral system model
Output	Complete and consistent model-integrated specification
Condition/limitation	-

The integration of models and textual specifications is based on a basic rule: a specification chapter is created for each element of the integrated model. Generally, model elements are related to the nodes and edges of a graph structure on which a model is based. Each model element is parsed and a specification skeleton is created for each model element. This allows, for example, a semi-formal specification based on requirements written in IBM Rational DOORS as well as the generation of models documented in MS Visio. The natural language specification that serves as input is hence supplemented by integrating the modeling elements from the model. In the subsequent specification task, a domain expert adds to and details this specification, supplementing the resulting specifi-

Basic rules for model integration

cation with conditional expressions to allow the models to be interpreted uniquely. Propositional logic is used to specify conditional structures systematically. Model integration has been applied to the example adaptive cruise control (ACC) from the automotive system cluster described in Chapter 2. Fig. 6-5 presents the behavioral model developed as a basis for the application of model integration. The textual requirements are outlined following states and state transitions of the model (Fig. 6-5). The model and textual requirements together form the model-integrated specification of the adaptive cruise control system.

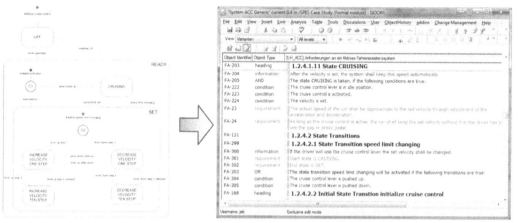

Fig. 6-5   *A behavior model given by a state machine is used to structure the chapters of the corresponding requirements specification*

Transformation of natural language requirements   The next step towards formalization targets the requirements within the chapters of the model-integrated specification, which are still described by natural language text. In the following method, requirements are formalized by transforming them into boilerplate specifications which are subsequently transformed into requirements specification language (RSL) specifications with a fixed syntax and semantics. As outlined in Section 6.2.1, boilerplates allow requirements to be captured consistently, which produces fewer spelling or grammatical mistakes, prevents ambiguities, and makes requirements easier to understand. This supports objectivity in the understanding of the problem space, thereby allowing for objectivity in validation results.

Tab. 6-3   *Formalization of natural language requirements*

SPES XT methodological building block	
ID	6.3
Name	Formalization of natural language requirements
Method	1) Transform the natural language requirements into semi-formal boilerplates. 2) Transform the boilerplates into formal RSL specifications.
Input	Model-integrated or natural language specification, domain knowledge
Output	Behavior specification
Condition/limitation	-

To create a semi-formal boilerplate specification (see also [Staalhane et al. 2010]), we need natural language requirements and domain knowledge. The chapters of the model-integrated specification, which is an outcome of Tab. 6-2, contain natural language requirements. The result of this formalization step is used as an input to generate a formal behavior specification. This combined formalization process, therefore, relates to the requirements viewpoint, where the boilerplate specification is located. The semi-formal boilerplates are then used for further formalization. While boilerplates enable lightweight checks — such as the consistency analysis between the terms used in the requirement and the domain knowledge — there is a need for more sophisticated analyses. Checking, for example, whether a set of requirements is consistent within itself requires an underlying semantic such as linear temporal logic (LTL). We therefore introduce the requirements specification language (RSL) (see also [Reinkemeier et al. 2011]), which consists of patterns such as:

Creation of semi-formal boilerplate specification

Whenever <event> occurs, <condition> holds during <interval>.

In the same way as for boilerplates, the requirements engineer selects the pattern and substitutes the placeholders with the entities of the requirement. Tab. 6-4 depicts the results of the two formalization steps applied to an adaptive cruise control system requirement from the automotive system cluster.

Tab. 6-4   *Formalization of a natural language requirement from the automotive*
                  system cluster example

Automotive system cluster example	
Adaptive cruise control system	
Natural language require-ment:	If the cruise control is deactivated and the cruise control lever is pulled, the last chosen speed setpoint should be adopted as the set speed.
Boilerplate specification: Boilerplate:	If the cruise control is deactivated and the cruise control lever is pulled, the cruise control system shall adopt the last chosen speed setpoint as the set speed. Pattern: if <state> and <event>, <system> shall <action>
Behavior specification: RSL pattern:	Whenever cruise_control_lever_pulled occurs under cruise_control_system == deactivated, set_speed == last_set_speed holds. Pattern: whenever <event> occurs under <condition>, <condition> holds

Several decisions are needed

To achieve this result, the engineer has to make several decisions, (e.g., splitting the requirement, selecting a suitable boilerplate/RSL pattern). If we look at the resulting boilerplate specification, even though it does not differ much from the original requirement, it reduces ambiguity by explicitly stating which system shall perform the intended action. This formalization process can be supported by tools (e.g., DODT [Farfeleder et al. 2011], Contract Editor) that comprise lists of predefined boilerplates and RSL patterns. The tools also check the terms entered against domain knowledge or for correct syntax.

Validation of the specification

Once the natural language requirements have been formalized, the resulting (semi-)formal specification enables the engineer to validate this specification with respect to consistency and completeness. This is especially helpful when requirements have been specified by different persons with different purposes. The principle of separation of concerns allows the engineer to focus on the aspect that is important for him. A disadvantage of this paradigm is that it is more difficult to identify dependencies between the different aspects. More specifically, a safety engineer may specify requirements that an engineer who is concerned with real-time properties might not be aware of. Therefore, the final building block of this section describes a multi-aspect validation that aims to capture dependencies between requirements of different *aspects* (e.g., safety and real time) to validate the requirements specification.

Tab. 6-5   *Multi-aspect validation*

SPES XT methodological building block	
ID	6.4
Name	Multi-aspect validation
Method	Validation of behavior specifications including dependencies between safety and real-time requirements.
Input	Behavior specification
Output	Analysis results
Condition/limitation	Behavior specification contains formalized requirements (as described in Tab. 6-4).

The multi-aspect analysis introduced is based on the behavior specifications of functional, logical, or technical components which are located in the corresponding viewpoints of the SPES XT modeling framework. Once the analysis has been performed, the results are fed back to the engineer, who can modify the artifacts accordingly. During the development of an adaptive cruise control system, subsystems such as the emergency braking assistant (EBA) are specified. Under the principle of the separation of concerns, one engineer specifies the real-time behavior of the system such that *the EBA shall start decelerating the vehicle within 30 ms of detecting an obstacle*. In addition, a safety engineer specifies the safety behavior of the EBA (e.g., *if the primary sensor for detecting an obstacle fails, the EBA shall switch to the secondary sensor*). Once both requirements have been formalized with boilerplates and RSL, an automated multi-aspect analysis identifies a dependency between both requirements due to the specification fragment *detecting an obstacle* that is part of both requirements. This may lead to a refinement of the real-time requirement because a sensor switch might invalidate the required timing property.

Multi-aspect analysis is based on behavior specifications

6.3.3   Validating Functional Adequacy through Simulation

In addition to consistency analyses (see Sections 6.3.1 and 6.3.2), which are mostly aimed at checking static properties of the specification, the dynamic behavior of a SUD embedded in its runtime-context must be validated. In this section, we describe a set of methodological building blocks that can be used to validate the

Dynamic behavior of a system at runtime must be validated

functional adequacy of system models in a systematic process by
generating and executing validation use cases in a (co-)simulation of
system and context models [Brandstetter et al. 2015a]. This method
is especially designed for validating models of a control software
against the physical processes that reside in its context. The process
is illustrated by applying it to the desalination plant example (see
Chapter 2 and [Brandstetter et al. 2015b]). The building blocks
address context modeling, requirements, and the functional viewpoint of the SPES XT modeling framework (see Chapter 3). The
method is a *validation technique,* of which the *objective* is to validate the functional adequacy of a *system model* against its operational context by means of (co-)simulation. Thus, this method is an
instantiation of the RQAF. The method is structured into several
methodological building blocks that together build a systematic
process: context modeling (Fig. 6-6), derivation of validation scenarios (Tab. 6-7), creation of executable system specifications (Tab.
6-8), and the execution of the actual (co-)simulation (Tab. 6-9). Fig.
6-6 gives an overview of the approach.

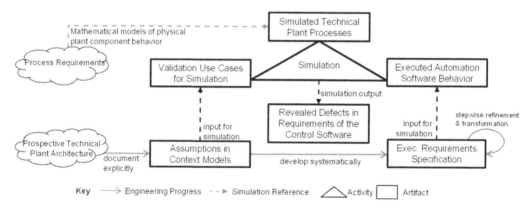

Fig. 6-6   *Overview of the validation technique*

Context models are created for validation

As a first step of this method, context models which serve as a
common reference for all subsequent models are created. From the
perspective of the automation software, the technical plant constitutes the operational context in this case. Therefore, we explicitly
document assumptions about the structural and functional characteristics as well as the behavior of technical plant components. To
this end, we apply the SPES XT context modeling framework (see
Chapter 4).

Tab. 6-6   *Context modeling*

SPES XT methodological building block	
ID	6.5
Name	Context modeling for simulative validation
Method	Document assumptions about the prospective system by means of context models.
Input	Natural language specification, engineering artifacts of external systems (e.g., piping and installation diagram of an automation plant)
Output	Structural, functional, and behavioral context models (see Chapter 4)
Condition/limitation	-

Structural operational context: Structural operational context models document the components of the operational environment that need to be considered in a simulation. Fig. 6-7 shows an example from the desalination plant from Chapter 2. A beach well (BW) consists of a pump, a bypass valve (BPV), a discharge valve (DV), and a pressure bypass. The automation software shares a number of interfaces with these beach well components — for example, interfaces to turn pumps on and off or to open and close valves. The user receives information about the plant from the automation software and can activate and deactivate single beach wells. The user can also set a required throughput for the beach wells.

Structural operational context

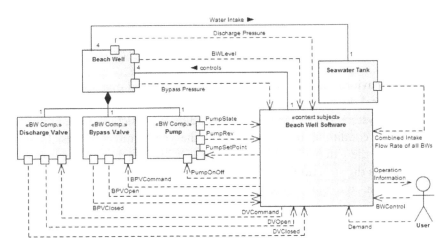

Fig. 6-7   *Structural operational context of the beach well control software*

Functional operational context: Functional operational context models are used to derive validation use cases — that is, interactions that may occur during operation which must be validated. Fig. 6-8 depicts the automation software, which sends control signals to the beach well functionality based on observation of the water flow rates from the seawater tank and the pump output pressure. The function *Start Beach Well* first activates the pump such that water pressure can build up. Once an output pressure threshold is reached, the function closes the bypass valve and opens the discharge valve so that water can flow into the seawater tank. Similarly, the function *Stop Beach Well* turns the pump off and opens the discharge and bypass valves respectively. The function *Balance Load* distributes the requested water flow across multiple beach wells.

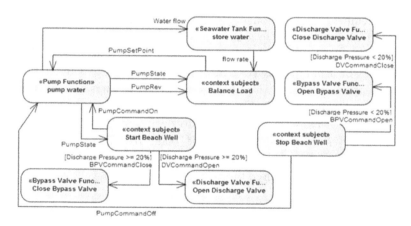

Fig. 6-8   *Functional operational context of the desalination plant beach wells*

Behavioral operational context: The behavioral operational context model documents the externally observable behavior of the simulated system and hence allows preconditions, postconditions, and success conditions to be derived for the validation use cases. In Fig. 6-9, the externally observable states of the beach well components from the structural operational context (i.e., bypass valve, discharge valve, and pump) are depicted as concurrent substates. Both the bypass valve and discharge valve can be either open or closed. The pump is off, in a startup condition, or in a shutdown condition. The guards on the transitions are conditions of the beach well function *Start Beach Well* specified in the functional operational context model. Based on these context models, we systematically develop validation scenarios and executable requirements descriptions as shown in the next sections.

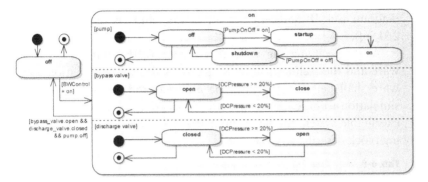

Fig. 6-9   *Behavioral operational context of the desalination plant beach wells*

All subsequent models that are created reference the context models. The first models are validation scenario models, which are later executed in a simulation environment. Validation scenarios are identified while setting up the context model. They contain sequences of steps that describe desired interactions between the system and its context as well as initial and final states of the system. For the definition of the initial states, the classification tree method is used [Grochtmann and Grimm 1993] (see Tab. 6-7).

Subsequent models

Tab. 6-7   *Generating validation scenarios*

SPES XT methodological building block	
ID	6.6
Name	Specification of validation scenarios based on context models
Method	Creation of validation scenarios using the classification tree method [Grochtmann and Grimm 1993]
Input	Context model, technical description (plant layout and technical data)
Output	Validation scenarios
Condition/limitation	-

In this method, the input domain of the system under validation is divided into different aspects (classifications) and possible values (classes). The functional operational context model defines an exchange of signals between the system and its environment. The signals are represented as classifications in a classification tree that illustrates the characteristics of the system under validation. Plausible signal values are classes. A validation scenario is specified by defining a particular combination of the classes of different classifications. A specification scenario can be illustrated by a scenario diagram (see [Brandstetter et al. 2015a] and [Brandstetter et al. 2015b]).

Execution of the validation scenario in a simulation

After specifying the validation scenarios, the automation software and the technical process against which the automation software shall be validated must be modeled and then coupled in a simulation to execute the validation scenarios. The creation of this executable system model is summarized in a methodological building block described in Tab. 6-8.

Tab. 6-8   *Creating executable system models*

SPES XT methodological building block	
ID	6.7
Name	Creating executable system models from context models
Method	1. Create an executable software model (see also [Vogelsang et al. 2014]). 1.1. Extract the system functions of the functional context model. 1.2. Extend the system functions with a behavior specification. 1.3. Derive an executable software model by composing all function specifications. 2. Create an executable process model. 2.1. Define the topology and parameters of the model based on technical diagrams (e.g., P&ID and technical data). 2.2. Derive the interface of the actors and sensors from the context models. 2.3. Enrich models with additional information (e.g., initial values).
Input	Static and functional operational context models, technical description (plant layout and technical data)
Output	Executable functional black-box model
Condition/limitation	-

Desired automation software behavior, system functions are defined

To derive an executable description for the desired automation software behavior, system functions are defined and described. Each function handles a subset of the input and output signals of the automation software that is specified in the structural context model. The system functions are described by a behavior specification. To that end, the function is either decomposed into further subfunctions or the behavior is specified by an executable behavior description (e.g., state machine, table specification, or code snippet). An executable model of the automation software behavior results from the parallel composition of behavior specifications of all (sub)functions (cf. [Broy 2010]). A function summarizes a set of requirements and formalizes them by means of an executable behavior description. For the specification of the functionality in our example, we define two functions: *Toggle Beach Well* and *Balance Load* (see Fig. 6-10). In this example, both functions are independent of each other. In an automation plant with a higher degree of automation — for example, when the *Balance Load* function may automatically start or stop a beach well — it might be necessary to add internal channels that model the communication between functions (cf. [Vogelsang et al. 2015]). Fig. 6-11 shows a state machine that describes the behavior of the *Toggle Beach Well* function.

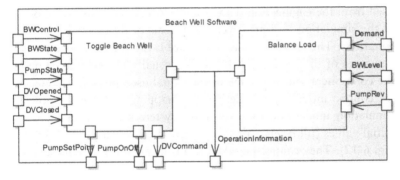

Fig. 6-10   *The automation software is structured into two functions*

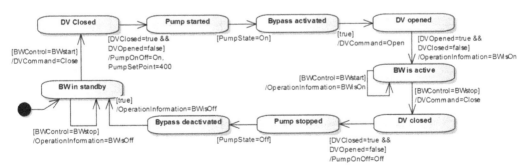

Fig. 6-11   *Behavior of the "Toggle Beach Well" function described by a state machine*

An executable behavior *specification is needed*

In addition to creating executable specifications of the automation software, we have to provide an executable behavior specification for the technical system that resides in the operational context of the automation software. In a technical system, an action triggered by an actor usually results in several items of feedback from the system back to the software (e.g., turning on a pump results in a change of several pressure sensors in the system). Therefore, it is hardly possible to validate the software without a model of the technical process to also simulate the second-order results of an initial action.

Set up the simulation *model in a component-* *based simulator*

Many plant-engineering tools use the object-oriented paradigm to structure engineering data. This means that for every component — for example, a beach well — technical information is stored for all disciplines involved (geometry, instrumentation, and interfaces to other objects, e.g., hydraulic connections). This information can be translated to set up the simulation model in a component-based simulator. The topology is derived from technical diagrams such as the P&ID diagram (see Chapter 2). The technical data can be re-used from the engineering data (see Fig. 6-12). The physical models of the simulation tool are given by systems of differential algebraic equations. The interface of the software is given by the sensors and actuators of the system. Their naming usually follows the name of the component and the name of the signal used in Fig. 6-7. Hence, the context model already defines the mapping between the process simulation model and the executable system specification. Finally, initial values (fill levels, salt concentrations, etc.) have to be set (see Fig. 6-12). They cannot be taken from the engineering data as they are not known and they might change for different validation scenarios.

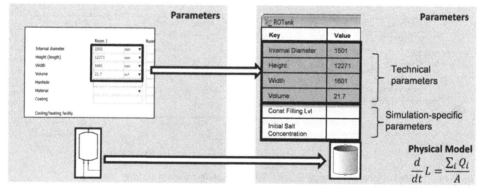

Fig. 6-12   *Automatic model generation: the engineering data of a beach well is transferred to a component-oriented simulator*

After creating validation scenarios and executable system models, we put the previous blocks together to perform a simulative validation. The block *Creating executable system models* (Tab. 6-8) contains the simulation model and the block *Generating validation scenarios* (Tab. 6-7) the initial conditions.

Tab. 6-9   *Co-simulation*

SPES XT methodological building block	
ID	6.8
Name	Co-simulation for early validation
Method	1) Initialize the simulation model with the models of the automation software and the technical system (Tab. 6-8). 2) Simulate the specified validation scenarios (Tab. 6-7).
Input	Validation scenarios, technical process model, automation software model
Output	Simulation results
Condition/limitation	Technical process model and automation software model must be executable.

In the simulation, the models communicate via a common signal pool. Each model, when added to the simulation, defines its interactions with other models by subscribing to the signal pool. The co-simulation method executes models in the same time domain with a deterministic response time. The knowledge about the execution time and run order of the models is necessary to achieve the correct validation results. It is important for the co-simulation to use the same time domain for all models. A configuration manager adds each imported model to the simulation. All necessary communication signals (signal pool) between the models are created from the interface description and will be mapped to each port of the models. While the simulation is running, all signals of the signal pool are available for simulation and monitoring and can be used for validating the functional adequacy by executing the specified validation scenario (see Fig. 6-13).

Communication between models

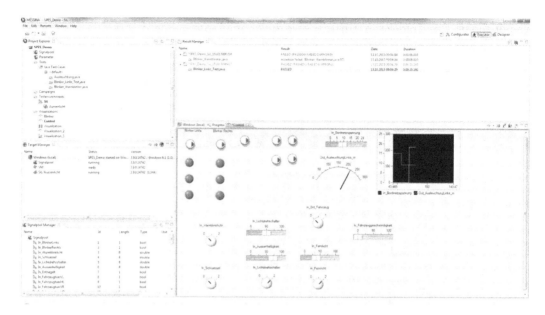

Fig. 6-13   *Visualization of a simulation run in MESSINA*

Timing plays a crucial role — Timing plays a crucial role in the simulative validation of embedded systems. In general, input/output behavior has some timing restrictions or the impact on the environment also depends on the specific points in time when resulting values are available. Usually, a set of functions is executed on a target system. They compete with each other for the execution time available on the electronic control unit (ECU) and for the communication time on the bus systems. Depending on the ECU and the functions, simple approaches such as a fixed, pre-calculated static schedule, a simple scheduler using only some interrupts, up to powerful operating systems can be used to distribute the available execution times to the different functions. An early simulation indicates whether the timing requirements are met by the system design being considered, and, for example, which execution time budgets and scheduling constraints have to be fulfilled by the subsequent implementation.

Simulation of executable specifications may focus on different aspects — A simulation of executable specifications may focus on different aspects. While a purely functional simulation estimates the output values based on the input values, the timing simulation predicts the influence of the intended platform — with scheduling, operating system, etc. — based on the functional values. The combination of both approaches allows the functional simulation to base the prediction on the correct input values and the timing simulation to use the correct calculated output values, which may influence the timing behavior. A timing simulation that considers functions allows timing properties to be integrated with a purely functional simulation. The signal pool is a data layer for data exchange across the components connected to the simulation framework. The specific values given to the system under validation and later back to the signal pool by the actuators depend heavily on the evaluation of the timing paths in the actual simulation run. The simulator ensures that the correct instance of data is given back at the correct time.

6.4 Summary

Maintaining the validity of engineering artifacts right from the start of a development project is a crucial activity because errors and failures in an early stage are propagated to subsequent activities. To enable an early and continuous validation of engineering artifacts, we identified four challenges that we addressed in this chapter. Since many engineering artifacts — especially in early phases — rely on textual descriptions, we need methods to increase the objectivity of validation results for natural language–based artifacts. In modern systems, there are a large number of artifacts to be considered. We need improved tool support and automation to make validation in early phases meaningful. Furthermore, the systems that we produce and analyze are becoming more and more complex. Therefore, we need methods that help us to handle the complexity of these systems. Finally, the systems we are dealing with are composites of mechanics, electronics, and software. Therefore, our methods need to consider the whole mechatronic system. In this chapter, we provided methodological building blocks for structuring and formalizing textual requirements, for analyzing the consistency of specifications, and for simulating system specifications for validating the resulting behavior. We considered functional aspects as well as timing and safety-related aspects of a system specification. In addition to the methodological building blocks, we provided the RQAF framework which is a blueprint for a structured validation approach. The methods and tools provided in this chapter are instances of the RQAF elements. Thus, the methods can be related to each other via the RQAF.

6.5   References

[Boehm 1981] B. Boehm: Software Engineering Economics. Prentice Hall, Englewood Cliffs, 1981.

[Brandstetter et al. 2015a] V. Brandstetter, A. Froese, B. Tenbergen, A. Vogelsang, J. C. Wehrstedt, T. Weyer: Early Validation of Control Software for Automation Plants on the Example of a Seawater Desalination Plant. In: Proceedings of CAISE 2015, pp. 189-196.

[Brandstetter et al. 2015b] V. Brandstetter, A. Froese, B. Tenbergen, A. Vogelsang, J. C. Wehrstedt, T. Weyer: Early Validation of Automation Plant Control Software for using Simulation Based on Assumption Modeling and Validation Use Cases. In: Complex Systems Informatics and Modeling Quarterly Issue 4, pp. 50-65.

[Broy 2010] M. Broy: Multifunctional software systems: Structured modeling and specification of functional requirements. In: Science of Computer Programming. Vol. 75, No.12, 2010.

[Daun et al. 2014] M. Daun, J. Brings, B. Tenbergen, T. Weyer: On the Model-Based Documentation of Knowledge Sources in the Engineering of Embedded Systems. In: Proceedings of the Fourth Workshop on the Future of the Development of Software-Intensive Embedded System Development (ENVISION2020), 2014, pp. 67-76.

[Daun et al. 2015] M. Daun, B. Tenbergen, J. Brings, T. Weyer: Documenting Assumptions about the Operational Context of Long-Living Collaborative Embedded Systems. In: Proceedings of the 2nd Collaborative Workshop on Evolution and Maintenance of Long-Living Software Systems (EMLS), 2015, pp. 115-117.

[Fagan 1976] M. E. Fagan: Design and Code Inspections to Reduce Errors in Program Development. IBM Systems Journal, Vol. 15, No. 3, 1976, pp. 258-287.

[Farfeleder et al. 2011] S. Farfeleder, T. Moser, A. Krall, T. Stålhane, H. Zojer, C. Panis: DODT: Increasing requirements formalism using domain ontologies for improved embedded systems development. In: Proceedings of the IEEE 14th International Symposium on Design and Diagnostics of Electronic Circuits & Systems (DDECS), 2011, pp. 271-274.

[Flynn and Warhurst 1994] D. Flynn, R. Warhurst: An Empirical Study of the Validation Process within Requirements Determination. In: Information Systems Journal, Vol. 4, No. 3, 2014, pp. 185-212.

[Gacitua et al. 2009] R. Gacitua, L. Ma, B. Nuseibeh, P. Piwek, A. de Roeck, M. Rouncefield, P. Sawyer, A. Willis, H. Yang: Making Tacit Requirements Explicit. In: Proceedings of the 2nd International Workshop on Managing Requirements Engineering Knowledge, 2009.

[Glinz and Fricker 2014] M. Glinz, S. A. Fricker: On shared understanding in software engineering: an essay. In: Computer Science - Research and Development. Springer, Berlin Heidelberg, 2014.

[Grochtmann and Grimm 1993] M. Grochtmann, K. Grimm: Classification Trees for Partition Testing. In: Software Testing, Verification & Reliability, Vol. 3, No. 2, 1993.

[ISO/IEC 25010] ISO/IEC: ISO/IEC 25010:2011-03 – Systems and software engineering - Systems and software Quality Requirements and Evaluation (SQuaRE) - System and software quality models. 2011.

[Lisagor et al. 2010] I. Lisagor, L. Sun, T. Kelly: The Illusion of Method: Challenges of Model-Based Safety Assessment. In: Proceedings of the 28th International System Safety Conference (ISSC), 2010.

[Reinkemeier et al. 2011] P. Reinkemeier, I. Stierand, P. Rehkop, S. Henkler: A pattern-based requirement specification language: Mapping automotive specific timing requirements. Software Engineering 2011 – Workshopband, 2011.

[Shull et al. 2000] F. Shull, I. Rus, V. Basili: How Perspective-Based Reading Can Improve Requirements Inspections. In: IEEE Computer, Vol. 33, 2000.

[Sikora et al. 2012] E. Sikora, B. Tenbergen, K. Pohl: Industry Needs and Research Directions in Requirements Engineering for Embedded Systems. In: Requirements Engineering, Vol. 17, No. 1, 2012, pp. 57-78.

[Staalhane et al. 2010] T. Staalhane, I. Omoronyiam, F. Reichenbach: Ontology guided requirements and safety analysis. In: Proceedings of the 6th International Conference on Safety of Industrial Automated Systems (SIAS 2010), 2010.

[Tenbergen et al. 2015] B. Tenbergen, T. Weyer, K. Pohl: Supporting the Validation of Adequacy in Requirements-Based Hazard Mitigations. In: Proceedings of the 21st International Working Conference on Requirements Engineering: Foundations for Software Quality (REFSQ), 2015, pp. 17-32.

[Vogelsang et al. 2014] A. Vogelsang, S. Eder, G. Hackenberg, M. Junker, S. Teufl: Supporting concurrent development of requirements and architecture: A model-based approach. In: Proceedings of the 2nd International Conference on Model-Driven Engineering and Software Development, 2014.

[Vogelsang et al. 2015] A. Vogelsang, H. Femmer, C. Winkler: Systematic Elicitation of Mode Models for Multifunctional Systems. In: Proceedings of the 23rd IEEE International Requirements Engineering Conference, 2015.

[Wiegers 2003] K. Wiegers: Software Requirements. Microsoft Press, Redmond, 2003.

Jens Oehlerking
Thomas Strathmann

7

Verification of Systems in Physical Contexts

One defining characteristic of embedded software systems is the tight interaction with the context of the system, which gives rise to real-time constraints for the software. In this chapter, we focus on one relevant class of software: systems interacting with a physical context. This class includes embedded software for different engineering domains, for instance, mechanical, electrical, or biological systems. Typically, the purpose of this class of embedded software is some form of control over the physical context. Therefore, to formulate system requirements and consider whether they are fulfilled, context models must be tightly integrated with models of the software itself.

This chapter describes a modeling formalism for both physical contexts and control software which enables this kind of integration: hybrid automata. We also describe how existing methods can be leveraged to perform verification on the closed-loop systems arising from the composition of context and software models. The goal here is to achieve early validation in the sense of Chapter 6 for this class of systems as well.

© Springer International Publishing AG 2016
K. Pohl et al. (eds.), *Advanced Model-Based Engineering of Embedded Systems*,
DOI 10.1007/978-3-319-48003-9_7

7.1   Introduction

Systems in physical contexts

In this chapter, we look at the interaction of embedded software systems with physical contexts. Typically, embedded software receives information from the context via sensor readings, performs computations based on that information, and influences the context via a number of actuators to achieve a predefined goal.

Physical requirements

Most importantly, system requirements are typically at least partially defined on the physical context. For example, a typical requirement for a cruise control system might take the form:

After activating the cruise control, bring the vehicle velocity to within +/- 10% of the velocity setpoint within 10 seconds

Here, the vehicle velocity is a state of the context which can only be manipulated indirectly — for example, via brake pressure or engine torque. Therefore, in order to apply model-based methods to this class of systems, models of the software and the context have to be dealt with in an integrated fashion.

Analyzing systems in a model-based fashion early in the development process carries a great benefit in terms of development costs. One class of analysis algorithms that can be applied to such models is that of *verification* algorithms, which seek to prove that a requirement is fulfilled or to provide counterexamples when it is not.

In the following, we differentiate between the task of deriving suitable models based on the formalism of *hybrid automata* and the task of leveraging these models for verification.

Generally, hybrid automaton models of a system and its context can be used not only for verification but also for documentation, calibration, or software architecture purposes. However, within the scope of this chapter, we restrict ourselves to the verification aspect.

In Section 7.2, we concentrate on the model aspects, introducing hybrid automata based on an example. In Section 7.3, we then describe the two methodological building blocks: modeling and verification.

Challenges in verifying embedded systems

Because context models cannot be assumed to exactly mirror the behavior of the physical system but merely approximate it, robust fulfillment of the requirements must be shown. This means that, even under reasonable variations of physical context behavior, the overall system should function properly.

These variations can carry various interpretations: sensor noise, disturbances not modeled (e.g., temperature changes), variations in physical parameters (e.g., the weight or size of a physical component deviating from what was specified), or conservatively restricted modeling errors (i.e., deviations between model behavior and actual physical behavior). Furthermore, some parameter values in the software are only instantiated late in the development process (e.g., on a system that is already deployed on a prototype vehicle) and are therefore unknown in the early stages of the process when model-based approaches are applied. Model-based verification techniques must also take these open parameters into account.

In addition, it is desirable to reuse the same piece of software for different physical system topologies (e.g., for a hybrid drive train as well as a combustion engine). This structural variability in the physics is then mirrored as structural variability in the software. Therefore, component-based design approaches are required where variability in the physical context also has to be taken into account.

In terms of verification algorithms, the main challenge here is to arrive at the correct abstraction levels of models for the given verification task. If the model is too high-level, it may not be possible to express a particular requirement. If it is too detailed, verification algorithms may fail due to their inherent complexity. Deriving suitable verification models must typically be done manually by an engineer familiar with both the domain and verification algorithms.

7.2 Extensions to the SPES Modeling Framework

As discussed above, the physical context of a software system is important for a large class of embedded systems. In terms of Chapter 4, this will typically be either a behavioral or structural operational context. Behavioral context models are necessary, amongst other things, for simulation, documentation, formal verification, or testing.

Structural context models are required to formulate structural variability in the context. Since structural variability in the physical context is often mirrored as structural variability in the software, these models are useful for deriving a suitable software structure covering a number of variants. This structuring of the software, based on physical connections or cause-effect chains, is usually conducted from the functional or logical viewpoint, as described in Chapter 3.

Physical context models will rarely be functional in the sense of Chapter 3, as meaningful functional behavior for closed-loop systems typically only arises in the composition of software and context.

Since models of the physical context exist during all stages of the design process, they are heterogeneous in terms of granularity level, viewpoint, and modeling paradigms. For example, a physical context model can exist as a number of mathematical equations, a graph model describing the physical structure of the system, a complex multibody simulation model, or a model abstracting from many details, to be used for controller design.

Moreover, models of the physical context are also regularly used as part of the software itself to improve the performance of the software in the closed loop via model-based control algorithms. This also requires tailored abstractions of the physical context, since computations must be performed in real time with limited capacities.

Therefore, the context models have to be treated as first class citizens, also carrying notions of abstraction, refinement, transformation, or composition comparable to software models. Hybrid automata are a good means for documenting aspects of the behavioral physical context as they are a tool-independent paradigm. In contrast to the other modeling paradigms used in this book, they can express the continuous evolution of physical systems as well as parameter uncertainties. In addition, they carry a rich theory of abstraction, (approximate) refinement, composition, model-based testing, and formal verification.

We will now introduce *hybrid automata* as a special class of behavioral context models that can capture physical context behavior so that it can be leveraged for verification methods. We do this using the ACC system version 1 as described in Chapter 2 as an example. We focus on a closed-loop model of the speed control together with a simple model of the vehicle's dynamics.

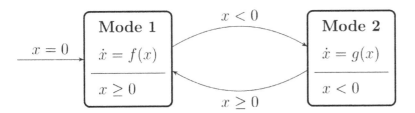

Fig. 7-1   *Example of a hybrid automaton*

A hybrid automaton (see Fig. 7-1) is defined via a set of variables and their derivatives with respect to time (denoted by a dot, e.g., $\dot{x}$). The continuous-time evolution of the state variables is described by means of ordinary differential equations and additional (in-)equations called *invariants*. The rounded rectangles represent the *modes* of the automaton, with the differential equations above the line and the invariants below. Note that automata theory common-ly refers to the definition of states rather than modes. However, in the domain of automotive engineering, a mode typically refers to some kind of global system state (e.g., startup, engine off, operating temperature reached). The arrows that connect the different modes are the discrete *jumps* of the automaton. If the *guard* inscribed on the arrow is true of the current valuation of the state variables, then the transition may be taken and the automaton changes from one mode to the next. For a more formal description of many of the relevant concepts, we refer to [Henzinger 1996].

A hybrid automaton can be seen as a refinement of the finite state machine-based behavioral operational context models as de-scribed in Section 4.2.2. Essentially, the (discrete) states of a state machine are enriched with continuous behavior in the form of dif-ferential equations which describe the physical laws influencing the software system under development (SUD). The resulting hybrid automaton model can then be used to consider requirements formu-lated on physical quantities, which are common in embedded sys-tems design. These physical quantities are typically modeled as vari-ables of the hybrid automaton.

Hybrid automata as a model for control software

For physical models, the mode changes typically represent changes in the structure of the physical context. Mode changes re-flect the different cause-effect chains possible in the context which can be made explicit by the use of hybrid automata.

Components and semantics of hybrid automata

However, hybrid automata can be used not only for modeling the context of a software system but also abstractions of the software itself. This is particularly useful for functional entities representing controllers that interact closely with a physical context. Control strategies are also typically described as differential equations early in the design process and only transformed into executable (discrete-time) software at a later design stage. The modes of a hybrid automaton would then represent different control strategies that can be applied depending on the situation — for example, mirroring the different cause-effect chains possible in the context. We will now illustrate these points by way of the ACC example from Chapter 2.

Fig. 7-2 shows the Simulink diagram that models the vehicle's dynamics. As input, it takes the acceleration set by the speed control and the brake pedal depression force. It outputs the vehicle's velocity, based on a lookup table (LUT) depending on the previous velocity value.

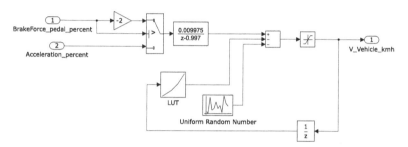

Fig. 7-2   *Simulink model of the vehicle dynamics*

Hybrid automaton for the physical context of the ACC function

Fig. 7-3 shows a hybrid automaton representation of the Simulink diagram. In the case of the vehicle dynamics model for the ACC system, the modes in the upper half of the automaton describe the different cause-effect chains in the physical context.

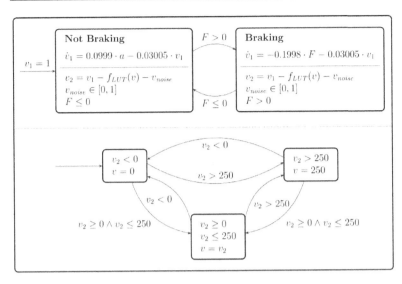

Fig. 7-3   *Hybrid automaton for the vehicle dynamics*

Either the brake system is engaged — at which point the vehicle's velocity v is influenced by the brake force F (mode *Braking*) — or it is not, so that the vehicle can be accelerated according to acceleration value a (mode *Not Braking*). The bottom half of the diagram shows a second parallel automaton which describes the velocity limitations that are assumed in the physics (velocity between 0 and 250 km/h).

The velocity controller for the *Not Braking* cause-effect chain, whose Simulink model is shown in Fig. 7-4, can be represented by the hybrid automaton in Fig. 7-5. The bottom half describes the limitations for the vehicle acceleration and the top half describes the different control modes: automatic acceleration (*Accelerate*), holding the velocity constant (*Stay*), and full control by the driver (*Pass Through*).

Hybrid automaton for ACC control software

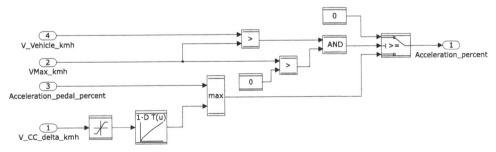

Fig. 7-4   *Simulink diagram of the velocity controller*

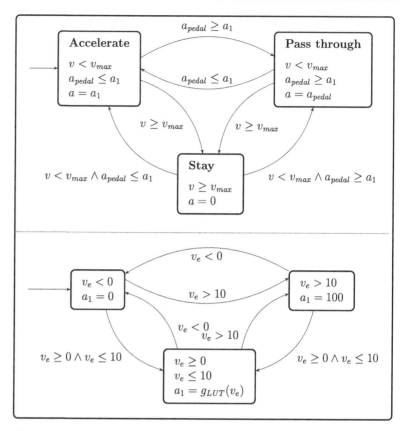

Fig. 7-5   *Hybrid automaton for the velocity controller*

Building a closed-loop model

In order to derive a closed-loop model of the velocity controller together with the model of the vehicle's dynamics, we have to compute the product of the hybrid automata in Fig. 7-3 and Fig. 7-5 (i.e., their composition). The interface between the two automata is the acceleration and velocity error given by the difference between the desired velocity v_{max} and the actual velocity v. This closed-loop model can then be used to show whether the velocity controller fulfills the requirements defined for the physical context.

Modeling parameter variations

The closed-loop model of the ACC may be subject to a number of parameter variations which can also be modeled as part of a hybrid automaton. In the model of the vehicle's dynamics, there are several physical parameters that vary depending on the vehicle for which the system is instantiated:

❑ The vehicle's mass
❑ The coefficients in the transfer function and the lookup table describing the vehicle dynamics

❑   The vehicle's maximum velocity

Furthermore, the sampling time and sampling jitter of the embedded software (which is not present in the Simulink model but could nevertheless be of interest for a parametric analysis) are also possible sources of parameter variations. These can also be modeled as part of hybrid automata [Frehse et al. 2014]. The model in Fig. 7-3 already accounts for sensor noise, as represented by the v_{noise} variable.

7.3   Methodological Building Blocks

This section discusses two methodological building blocks, namely the modeling of a system as a hybrid automaton and the use of formal verification techniques to ascertain whether the SUD satisfies its requirements.

The methodologies described here are geared towards explicit modeling of physical system contexts and possibly also their interaction with the software. These models can then be leveraged for testing, documentation, controller design, calibration, verification, and validation.

The first methodological building block describes the modeling paradigm of hybrid automata, which can be used to represent physical contexts as well as control software. The second methodological building block is the application of formal verification and validation techniques to such a model. In particular, the use of models consisting of both physical and software behavior allows consideration of requirements formulated in terms of physical quantities.

Modeling as a Hybrid Automaton

This methodological building block creates a hybrid automaton model amenable to formal verification given the functional and logical structure of the SUD. As the granularity level for such a model is dependent on its intended use, this is a manual modeling step.

The hybrid automaton model can represent the physical context of the embedded software (containing equations describing the relevant physical processes) or parts of the software system itself. In the latter case, the model would consist of continuous-time representations of the control algorithms. For closed-loop analysis, a composite model containing representations of software and context is required.

Acceptable limitations on the parameter uncertainties should also be encoded in the hybrid automaton and must therefore be established beforehand. To deal with variability, hybrid automata can also be modeled in a compositional fashion, enabling reuse of partial models.

Tab. 7-1   *Modeling as a hybrid automaton*

SPES XT methodological building block	
ID	7.1
Name	Modeling of a system as a hybrid automaton
Method	Equations for the system (e.g., from a simulation model) are manually translated into a hybrid automaton that captures the behavior necessary for a formal verification.
Input	Equations for physical context and/or equation representations of control algorithms, limitations on parameters
Output	Hybrid automaton
Limitation	Applicable to control software which is tightly coupled with a physical context

Parametric Verification of Hybrid Automata

This method addresses the problem of providing strong evidence that a system fulfills its functional requirements. Given a hybrid automaton that incorporates parameter variations and a formalized requirement, a number of verification methods may be able to check whether the system satisfies the requirement while in some cases providing counterexamples or safety margins.

A number of academic methods are available that can be used to check requirements on hybrid automaton models. These methods fall into two different classes: simulation-based falsification methods and formal verification methods.

Simulation-based falsification methods do not require a formal hybrid automaton model. Instead, they can work directly on simulation models. In general, given a requirement in some form of temporal logic, these methods leverage mathematical methods, trying to falsify the requirement. This means that the algorithms are geared towards identifying the "worst possible" simulation run with respect to the given requirements. If the method succeeds in producing a run which violates the requirement, it is falsified and the counterexample can be used to refine either the requirement or the model. If it does not, then no formal guarantees about the fulfillment of the requirement can be made, as full simulation coverage for continuous-valued systems is impossible to achieve in general.

Fig. 7-6 shows a counterexample which falsifies a simulation run. The dashed line depicts the trajectory of the reference speed which starts at 100 km/h and is changed to 110 km/h. The solid line depicts the vehicle's actual velocity. Simulation can automatically find this run which violates the requirement that a change in the reference speed starting from a stable cruising speed leads to a new stable cruising speed only deviating at most by 5 km/h from the reference value within less than one second.

Tab. 7-2   *Parametric verification of hybrid automata*

SPES XT methodological building block	
ID	7.2
Name	Parametric verification of hybrid automata
Method	Simulation-based falsification or formal verification of requirements for hybrid automata
Input	Depending on the tool, either a hybrid automaton or simulation model and a formal requirement
Output	Evidence that the requirement holds or is violated
Condition	• The model must be of a class that is supported by a verification tool. • The requirement must be given in a formalism supported by a verification tool.
Limitation	Depending on a number of properties of the hybrid automaton and the formal requirement, the method might not produce a result. This is due to theoretical reasons of decidability as well as technical limitations of currently available tools.

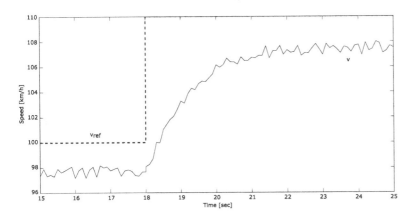

Fig. 7-6   *Example of a falsifying system run*

In contrast, formal verification methods aim to provide formal proof of the correctness of the requirement for the given model. Because this proof cannot be provided by simulation alone, a hybrid automaton model is required. In this case, for instance, "flow pipes," (sets of potentially reachable system states over time) can be computed and then related to the requirements.

Academic tools available for simulation-based falsification include *S-TaLiRo* [Nghiem et al. 2010] and *Breach* [Donzé 2010]. For the formal verification of hybrid automaton properties, tools such as *SpaceEx* [Frehse at al. 2011], *Flow** [Chen at al. 2013], or *iSAT-ODE* [Eggers et al. 2012] are available.

7.4   Summary

In this chapter, we presented hybrid automata — a modeling paradigm for physically controlled systems that can be used for both context modeling and modeling control software. Numerous academic tools can analyze this class of models, either based on formal methods or on simulation. Based on the ACC example, we showed how verification methods implemented on these tools can be applied and we discussed conditions under which these methods and tools are applicable.

The modeling paradigm and the verification methods are generally domain-independent and can be applied to mechatronic, electrical, and biological systems, amongst others, as long as the software and context can be represented by differential equations embedded in discrete logic. It is particularly useful for closed-loop control software characterized by a tight interaction between control software and the physical system context.

In the SPES XT modeling framework, hybrid automata are a specialization of behavioral context models for physical contexts, as well as a modeling language for control software.

With the modeling-based techniques described in this chapter, design errors can potentially be found earlier in the development process, enabling early validation in the sense of Chapter 6 also for physically controlled systems.

7.5   References

[Chen at al. 2013] X. Chen; E. Ábrahám; S. Sankaranarayanan: Flow*: An Analyzer for Non-Linear Hybrid Systems. In: N. Sharygina, H. Veith: Proceedings of the

25th International Conference, CAV 2013, Saint Petersburg, Russia, July 13-19, 2013, pp. 258-263.

[Donzé 2010] A. Donzé: Breach, A Toolbox for Verification and Parameter Synthesis of Hybrid Systems. In: T. Touili, B. Cook, P. Jackson: Proceedings of the 22nd International Conference, CAV 2010, Edinburgh, UK, July 15-19, 2010, pp. 167-170.

[Eggers et al. 2012] A. Eggers; N. Ramdani; N. Nedialkov; M. Fränzle: Improving the SAT modulo ODE approach to hybrid systems analysis by combining different enclosure methods. In: Software & Systems Modeling, Vol. 14, No. 1, 2012, pp. 121-148.

[Frehse at al. 2011] G. Frehse, C. Le Guernic, A. Donzé, S. Cotton, R. Ray, O. Lebeltel, R. Ripado, A. Girard, T. Dang, O. Maler: SpaceEx: Scalable Verification of Hybrid Systems. In: G. Gopalakrishnan, S. Qadeer: Proceedings of the 23rd International Conference, CAV 2011, Snowbird, UT, USA, July 14-20, 2011, pp. 379-395.

[Frehse et al. 2014] G. Frehse, A. Hamann, S. Quinton, M. Woehrle: Formal Analysis of Timing Effects on Closed-Loop Properties of Control Software. Real-Time Systems Symposium (RTSS), 2014 IEEE, 2014, pp. 53-62.

[Henzinger 1996] T. A. Henzinger: The Theory of Hybrid Automata. Proceedings of the 11th Annual IEEE Symposium on Logic in Computer Science (LICS), 1996.

[Nghiem et al. 2010] T. Nghiem, S. Sankaranarayanan, G. Fainekos, F. Ivancić, A. Gupta, G. J. Pappas: Monte-Carlo Techniques for Falsification of Temporal Properties of Non-Linear Hybrid Systems. In: Proceedings of the 13th ACM International Conference on Hybrid Systems: Computation and Control (HSCC '10), ACM. New York, 2010, pp. 211-220.

Karsten Albers
Stefan Beck
Matthias Büker
Marian Daun
John MacGregor
Andrea Salmon
Raphael Weber
Thorsten Weyer

8

System Function Networks

Embedded systems are often highly integrated into a network of systems. To increase synergies, reduce code redundancies, and to support reuse, the different systems in this network provide conceptual functions for use by other systems. This chapter deals with the challenges resulting from this type of networked approach and provides essential solution concepts for obtaining benefits from the creation and analysis of the model-based documentation of the functional design of such conceptual system function networks. In particular, this chapter explains:

- ❏ *The documentation format of the functional design of networked embedded systems, consisting of static/structural function networks, the functional behavior, and the timely execution order*
- ❏ *Analysis methods to aid validation and verification, as well as an optimal partitioning and deployment of networked embedded systems*
- ❏ *Construction methods to integrate the consistent creation of the functional design into embedded systems' development processes*

© Springer International Publishing AG 2016   119
K. Pohl et al. (eds.), *Advanced Model-Based Engineering of Embedded Systems*,
DOI 10.1007/978-3-319-48003-9_8

8.1   Introduction

System and function networks

In the past, embedded systems were developed as compositions of sensors, actuators, a computation platform, and the software function. Nowadays embedded systems are not designed to provide one functionality exclusively. Instead, new functionality is defined by the interaction of several systems and software functions. Software systems, despite being independent, are combined with other systems to form a network of systems and are implemented on a generalized execution platform. A system itself may consist of several functions which are, in turn, connected and interacting. In addition, each system can use not only its own functions but also functions provided by neighboring embedded systems. As a consequence, the system is part of the overall function network of the supersystem (i.e., the vehicle or the aircraft).

Function-centered engineering

In function-centered engineering processes the main point of reference are the system's functions, regardless of whether they will be implemented in hardware or software. Therefore, the functional design plays a central role; it defines the system functions to be implemented, functions in the system's context which are to be used, the intended behavior of each single function, the behavior resulting from interaction between functions, as well as safety-relevant quality properties such as real-time behavior. The functional design is constructed to support reuse, avoid code redundancies, and minimize the number of electronic control units. Hence, the functional design of a system under development uses functions which are provided by and under the control of neighboring systems.

8.1.1   Challenges regarding System Function Networks

Challenges in function-centered engineering

While some initial steps have been taken in the development of continuous model-based function-centered engineering processes – for example, by the definition of the AUTOSAR[2] standard in the automotive industry – several methodological questions remain unanswered:

---

[2] AUTOSAR - AUTomotive Open System ARchitecture, see http://www.autosar.org

❑   How can model-based engineering methods be applied to support continuous engineering of such system function networks?

❑   How can the resulting behavior of such function networks be validated and verified? For example, how can we ensure that interactions between systems and functions in such networks do not result in conflicts and in undesired emergent behavior?

❑   How can we ensure that systems consisting of such distributed function networks meet not only the required functional behavior but also the required properties in terms of performance, timing, and reliability?

During daily work, providing evidences for system properties mentioned above consumes a large part of the development time. In addition, the increasing complexity of such networks results in qualitative problems. Therefore, automated engineering and analysis techniques can increase the efficiency and effectiveness of the engineering process.

Automated techniques and modeling notations

8.1.2   Functional Design

This section introduces extensions to the SPES modeling framework to allow advanced documentation and analysis of the functional design in the SPES XT modeling framework. The functional viewpoint (see [Vogelsang et al. 2012]) of the SPES modeling framework is extended with diagram types to explicitly specify interconnectivity among functions of different systems in one function network. In addition, to aid function-centered engineering, methodological building blocks that aim to support the engineers with semiautomated answers to existing challenges are defined.

Metamodel extensions and method definition

In order to integrate and apply the different techniques described in this chapter in a seamless manner, we define an overall framework which guides the systems engineer in how and when to use the different methodological building blocks. The framework (see Fig. 8-1) gives an overview of how to integrate the techniques into one engineering process. The interfaces of each technique are well defined and are integrated into the SPES modeling framework described in Chapter 3. All techniques are based on the same function-specific metamodel extension detailed in Section 8.2.

As depicted, the techniques allow an automated synthesis of an initial functional design consisting of function networks and timing paths. Function networks may be analyzed when any emergent behavior occurs. Formal function network contracts and interfaces can also be derived from the model-based requirements specifica-

Functional design and requirements

tion. With further development and refinement of the functional design, the requirements and the function networks may become inconsistent. To avoid such disruptions in the development process, the synthesis also addresses the consistent development of behavioral requirements and functional design. More detailed descriptions of the methods provided can be found in Section 8.3.

Validation and verification

To ensure correct and compliant development, several methods for validation and verification purposes are provided. For example, function network and timing paths are used to verify the compliance of the timing behavior and the correctness of composition. The correct composition of the function network can also be verified by a virtual integration test considering multiple aspects (e.g., real time and safety).

Partitioning and deployment

Finally, function networks with interfaces and formal contracts are automatically partitioned into task structures. The correctness of this partitioning can also be verified. The final result of the partitioning process step serves as input to the optimal deployment challenge covered in Chapter 7.

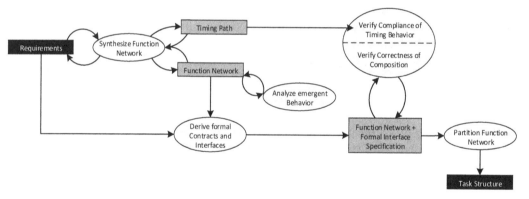

Fig. 8-1   *Methodological framework for function-centered engineering*

8.2   Extensions to the SPES Modeling Framework

Document functions, behavior, dependencies

This section introduces the function network extension to the SPES modeling framework. The extension allows the documentation of functions, their interaction and dependencies, their behavior, and their timely order. The documentation of these conceptual functions allows early validation activities (see Chapter 6) and optimized deployment (see Chapter 9) and benefits from the SPES XT context modeling framework (see Chapter 4). For separation of concerns, the function network extension is divided into a static/structural description of the function network that arises between different networked systems, the functional behavior of the function network arising from the interaction between each single function's behavior, and timing paths defining the specified interaction sequences within the function network.

8.2.1   Static/Structural Function Networks

As already mentioned, networked embedded systems are closely integrated, thereby sharing each other's functions to generate highly synergetic effects. Furthermore, functional properties exist not only in a single system's function but also within the interaction of multiple functions. Fig. 8-2 shows a high-level function network diagram for the adaptive exterior lighting system (ELS) from the automotive system cluster from Chapter 2. The function network diagram introduces the main functions of the system. In addition, it depicts the interactions and dependencies between the system functions of the system under development (SUD, i.e. the ELS) as well as with context functions provided by other systems of the vehicle. Note that we use excerpts from an entire system specification of the ELS which can be found in [Föcker et al. 2015].

Function sharing

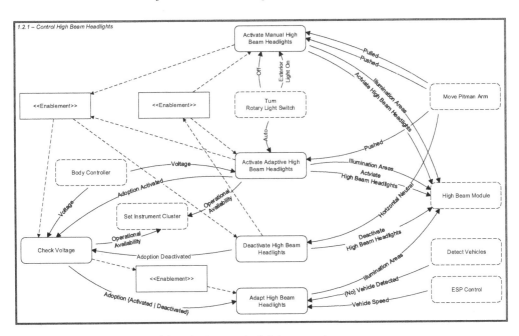

Fig. 8-2   *Function network of the adaptive exterior lighting control*

Basic modeling concepts The static/structural function network diagram consists of multiple modeling elements:

- ❑ *System functions (rounded rectangles with solid lines)* define conceptual functions that are provided by the SUD and can be used by other functions.
- ❑ *Context functions (rounded rectangles with dashed lines)* are functions that are not part of the SUD but are used by the SUD or that constrain the conceptual functions of the SUD. Note that context functions are related to the functional operational context from Chapter 4.
- ❑ *Interactions (solid arrows)* represent conceptual communication between different functions of the function network.
- ❑ *Dependencies (rectangles and dashed arrows)* document logical constraints between systems functions (e.g., to deactivate the high-beam headlights of the ELS, these must first have been activated manually or automatically at system startup), which might result in design constraints (e.g., acceleration and deceleration can be deployed to the same single core electronic control unit since functions will not need the core in parallel).

Advanced modeling concepts Furthermore, it might be necessary to depict the function network at a greater level of detail. In doing so, functions might be decomposed into more detailed subfunctions, and the focus of the diagram might be shifted to a single high-level function. To increase the expressiveness, additional modeling elements can be used, such as:

- ❑ *Ports* capsule inputs and outputs of a function in a specific entity. Thereby, each function can have multiple ports, each defining multiple inputs and outputs. The separation into ports allows a more detailed analysis and supports the specific deployment of functions to electronic control units (i.e., a function port is mapped to a hardware component port).
- ❑ *Messages* refine interactions that basically consist of message exchange (e.g., communication of a specific value).
- ❑ *Signals* refine interactions that basically consist of single electric signals (e.g., function calls or trigger conditions exchanged).

8.2.2   Function Network Behavior

Defining functional behavior While the function network diagrams document the static/structural relationships between conceptual functions and networked systems, the behavior of a function network defines the dynamic properties of a networked system. The functional behavior results from the interaction of each single function's behavior across the borders defined in the function network diagrams. Fig. 8-3 depicts a function behavior diagram for the function *Activate Adaptive High-Beam Headlights* from the function network diagram of the ELS from Fig. 8-2. The diagram in Fig. 8-3 uses the interface automata notation (cf. [Alfaro and Henzinger 2001]). The outer circle defines inputs to and outputs from the function. The inner state machine defines the accepted processing sequences of inputs and outputs.

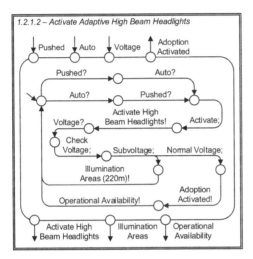

Fig. 8-3   *Behavior diagram for the function "Activate Adaptive High-Beam Headlights" from the function network in Fig. 8-2*

The functional behavior diagram's elements are closely related to the function network diagram:

Behavior modeling concepts

❏ *The diagram itself* is related to *the function* or a *single function's port* in the function network whose behavior is documented.
❏ *States* can be considered as decision points depending on external stimuli (the sending of inputs, receiving of outputs). The function's behavior changes with regard to the interacting functions from the function network.
❏ *Inputs* are related to *interactions, messages, or signals* depicted in the function network and impact the function (port) under investigation.
❏ *Outputs* are related to *interactions, messages, or signals* depicted in the function network and are impacted by the function (port) under investigation.

❑ *The execution order* is constrained by the *dependencies* defined in the function network.

8.2.3 Timing Paths

Information flow and processing

A vital task in embedded development is describing the flow of information and the information processing. The processing behavior describes several sequentially performed steps involving different functions and interactions between them. We use the concept of timing paths to describe such information flow and processing in a distributed system. A timing path consists of a sequence of steps which are traversed in their specific order during information processing. These steps can involve functions or interactions and are part of the functional network. In the underlying model, the dependencies between consecutive steps can be given by a processor–consumer or an input–output relationship.

In many cases, for an end-to-end timing path, the complete chain between the capturing of sensor values and the resulting commands to actuators is covered by one timing path.

An example of a timing path can be found in the adaptive light case example depicted in Fig. 8-4. It covers the functionality of reducing the high beam when an oncoming vehicle is approaching. The timing path describes the complete chain from the camera which captures the oncoming traffic, through the processing of the captured picture, the decisions about the reactions required, up to the actuator which reduces the light. The timing path has to cover steps such as image processing, object detection, object classification, decision on light reduction, processing of light reduction, as well as the messages between these steps.

Timing path as analysis artifact

A timing path is defined by the combination of the functional design and particular timing requirements. In addition, a timing path can be documented explicitly as an analysis artifact. Timing requirements, such as end-to-end response times, can be traced to these timing path artifacts which depict the data processing that must adhere to the requirement. In the example, the requirement *If a camera detects an oncoming vehicle, the high beam will be reduced to lower beam within 0.5 seconds* is in fact a worst-case response time requirement. This requirement consists of two separate parts, one concerning the frequency of the captured pictures and one concerning the maximum time allowed for processing one picture. The interaction of both parts must ensure the 0.5 second reaction time.

Natural
Language
Requirement

AL-16: If a camera detects an oncoming vehicle, the
high-beam will **be reduced** to low-beam **within 0.5 s**

Timing Path
(Requirements

Function   Network
Diagram   (Functional
Viewpoint)

Architecture
(Logical Viewpoint)

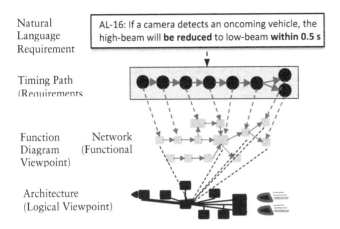

Fig. 8-4   *Example timing path*

The timing path relates the requirement and the functions designed to fulfill the requirement, allowing the functions to be refined and changed without changing the requirement. This separation supports the traceability of the requirements during the system development process. The refinement of the functions can be reflected in the timing path, resulting in a hierarchical structure of timing paths and timing sub-paths.

One function/message can appear several times in the same timing path. As the order of the steps within the timing path is fixed, several steps referencing the same function will not lead to infinite loops. A timing path artifact can include forks and joins. Forks model the multiple use of the same data for several different data processing steps; joins represent a calculation based on the several different inputs. The timing path allows a detailed and comprehensive investigation of the timing behavior and the fulfillment of the respective timing requirements by simulations and analysis methods.

Fig. 8-5 depicts an example timing path from the function network of the ELS in the notation format of ITU message sequence charts (MSCs) [ITU Z.120]. In detail, the timing path defines possible execution orders for the activation of the high-beam headlights. To represent and map a timing path, it is important to know:

- ❑ *Functions* that are involved within the timing path
- ❑ *Interactions* which are exchanged between the functions to execute the timing path

❏ *Execution orders* of the interactions that are allowed to reach the timing path's goal

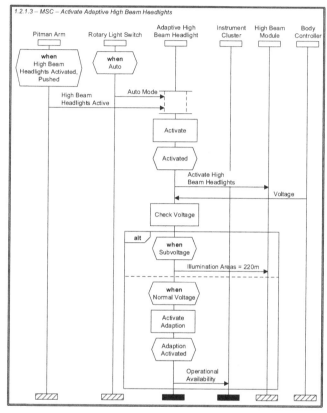

Fig. 8-5   *Diagrammatic representation of a timing path in the function network of the automotive system cluster*

8.3   Methodological Process Building Blocks

Model transformation and validation

To support function-centered engineering of embedded systems, this section presents several techniques which aid the engineering processes. In particular, it illustrates techniques for the continuous integration of the functional design into a model-driven engineering process demonstrating methods for the synthesis of the functional design from the requirements and for the generation of subsequent artifacts (see Section 8.3.1). In addition, this section presents validation techniques to ensure the correctness of the functional design. As well as approaches for dealing with the behavioral correctness of the functional design (see Section 8.3.2), this section presents techniques which aim at the development of a correct timely behavior in the composition of multiple functions (see Section 8.3.3).

8.3.1   Integration of Function Networks into Continuous Model-Driven Engineering

This section introduces a technique for synthesizing an initial functional design from the behavioral requirements as well as a technique to support the partitioning and deployment of functions. As Fig. 8-6 shows, these methods target the evolution of artifacts of the SPES XT modeling framework. Requirements can be synthesized into an initial functional design and logical components are partitioned from functions.

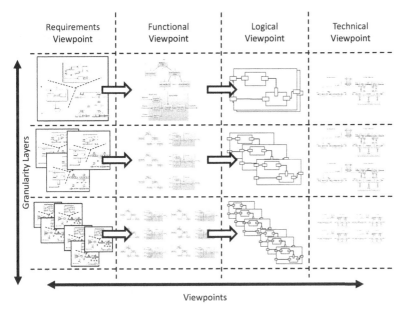

Fig. 8-6   *Seamless integration in the SPES XT modeling framework*

Initial Creation of the Functional Design

To support efficient model-based engineering with a high degree of consistency between the different development artifacts, initial artifacts for the functional design can be generated in a fully automated manner. Thereby, the function network diagram and the function behavior diagrams are consistently built by model transformations from the behavioral requirements. Tab. 8-1 briefly sketches the

Seamless and consistent engineering

proposed technique for synthesizing the functional design from the requirements specification.

The technique addresses the need for continuous highly auto-mated engineering support in the development of embedded sys-tems. In particular, it aids the function-centered engineering of em-bedded systems in so far as it allows automated generation of an initial version of the functional design which is consistent with the specified requirements.

The method generates initial versions of a function network dia-gram consisting of system functions representing the functionality of the system, the related context functions, and a function behavior diagram. Based on these initial artifacts, the engineers refine and detail the functional design in a manual engineering process.

Tab. 8-1   *Synthesis of an initial functional design*

SPES XT methodological building block	
ID	8.1
Name	Synthesis of an initial functional design
Method	The method synthesizes an initial functional design from a given behavioral requirements specification. To do so, the method relies on common model transformation algo-rithms. The detailed description of the method can be found in [Daun et al. 2014b].
Input	Interaction-based specification of the behavioral require-ments in MSC notation
Output	Function network diagram; function behavior diagram for the overall behavior
Condition/limitation	-

Application of the technique

For example, a simple basic message sequence chart ([ITU Z.120]) from the requirements specification (as depicted in Fig. 8-7) defines the automatic activation process for a vehicle's low-beam head-lights. After transformation by use of the method and the refine-ment by the engineers, a functional design is created which is con-sistent with the specified behavior from Fig. 8-7.

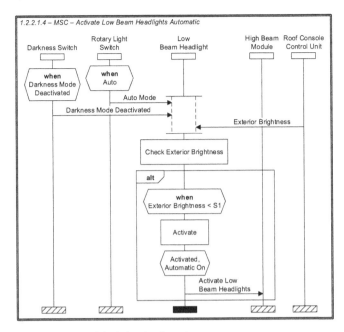

Fig. 8-7   *Excerpt of the behavioral requirements*

Fig. 8-8 shows the relevant part of the synthesized function network diagram resulting from the requirements for the automatic activation of the low-beam headlights. In addition, the behavior of the function *Activate Low-Beam Headlights Automatic* is shown. As we can see, this behavior is in accordance with the defined behavior from Fig. 8-7.

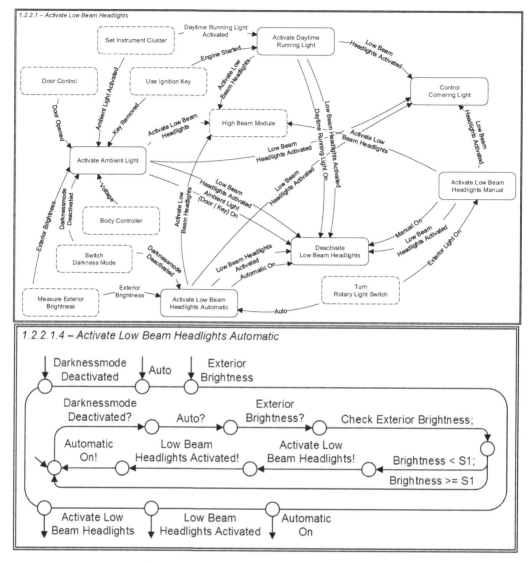

Fig. 8-8   *Functional design for the function "Activate Low-Beam Headlights Automatic" resulting from synthesis of the behavioral requirements*

Function Partitioning and Deployment

Logical architecture and components

As a further step towards the final system architecture, the functions identified in the functional design are partitioned into logical components in the logical viewpoint (LVP). Several functions of the functional design are clustered and partitioned to build one logical component. The logical architecture is intended to determine which functions are realized in which logical component and to identify and define interfaces between these components. The definition of logical components is a vital task in defining an optimized deployment of software components to hardware building blocks later on (see Chapter 9).

The function partitioning technique helps us to decide how to combine single functions into logical components. To do so, the concept of traceability between the functional design and the resulting logical architecture is of vital importance. To achieve traceability during the transition from the functional to the logical viewpoint, functions have to be mapped to those logical components that realize the respective function. In the simplest case, this is a 1:1 mapping relationship where each function is represented by a single atomic logical component. If one component should realize a set of different functions, all these functions have to be mapped to this component, ending in n:1 mapping relationships.

Achieve traceability

Although the mapping of one function to more than one logical component (1 : n mapping relationship, or even m:n mapping relationships) is possible, in general, such mapping relationships are problematic because it is often not obvious which parts of the functions are mapped to which logical component. Thus, the need for such a mapping relationship is often an indicator that the respective function should first be further decomposed into subfunctions.

1:n mapping indicates need for decomposition

Another important aspect is the explicit modeling of communication entities in terms of signals representing a data exchange between tasks. By defining signals, we can define what volume of data is exchanged at which rate and in which time frame. Although communication may already be modeled explicitly in the functional design by means of the message concept, further information regarding the partitioning and deployment may often be necessary. For example, in many cases, explicit communication entities are not necessary from a functional point of view. In addition, the messages of the functional design are not necessarily identical to the bus messages of the technical architecture because the former concept is a much more abstract and non-technical way of modeling communication.

Explicit documentation of communications

The design step of partitioning functions into components may also be supported by an automated optimization technique denoted as *Task Creation,* see [Büker et al. 2011] and [Büker 2013]. This technique subsequently merges functions following an optimization metric [Büker et al. 2013]. For further details on the optimization method, please refer to Chapter 9.

Component creation

8.3.2   Validation and Analysis of Functional Behavior

Validation of the Refined Functional Design

Taking the stakeholders into account Since the refined and detailed version of the functional design can easily become inconsistent with the original requirements specification, the functional design must be validated. For once, the engineering process of the functional design itself may lead to an increase in the knowledge about the system to be implemented. Hence, due to the fact that more information is elicited and documented than originates from the behavioral requirements, these initial behavioral requirements might no longer be considered correct and both artifacts must be validated closely against the actual stakeholder intentions (cf. [Daun et al. 2015c]).

Automated review model generation To achieve this goal, we propose a manual validation technique (cf. [Daun et al. 2014b]) which is based on the automated creation of a dedicated review model and the automated processing of the manually corrected review model to generate up-to-date versions of behavioral requirements and functional design. The process building block in Tab. 8-2 briefly summarizes the automated generation of the review model; the process building block in Tab. 8-3 summarizes the manual review process and the automated creation of up-to-date development artifacts.

Manual review of the review model The method may be used at any time during engineering to validate behavioral requirements and functional design. Process building block 8.2 describes the generation of a dedicated review model which will then be subject to manual inspections. The review model is documented in the same notation as the behavioral requirements to make it easier to check the stakeholder intentions.

The method described in Tab. 8-3 is helpful in the development of high-quality safety-critical systems and software. In particular, the method allows early validation of engineering artifacts to reduce costs and prevent an increase in defects in later development phases.

Tab. 8-2   *Method description for the automated creation of the review model*

SPES XT methodological building block	
ID	8.2
Name	Creation of a review model
Method	According to the detailed description in [Daun et al. 2015c], the method can be summarized as follows: 1. Creation of the overall behavioral model of the functional design: the function behavior diagrams for each function are combined to create one function behavior diagram. This function behavior diagram is then transformed into an MSC-like overall behavioral model. 2. Detection of basic message sequence charts (bMSCs), which are consistently refined by the functional design. 3. Detection of unrefineable bMSCs: unrefineable bMSCs of the behavioral requirements are detected, marked as unrefineable, and copied into the review model. 4. Detection of interaction sequences contained in the overall behavioral model of the functional design that are not specified in the behavioral requirements: these sequences are also transformed into bMSCs and marked specifically. 5. Enhancement of high-level message sequence charts (hMSC): Parts of newly created bMSCs that are already represented by existing bMSCs are detected and discarded. The hMSC is enhanced such that new bMSCs are inserted at the correct positions.
Input	MSC specification of the behavioral requirements; functional design consisting of function network diagram and corresponding function behavior diagrams
Output	Review model in MSC notation
Condition/ limitation	Behavioral requirements and functional design must be traceable either by explicit traceability links or by means of name identity

Tab. 8-3   *Method description for the review and deficiency correction*

SPES XT methodological building block	
ID	8.3
Name	Validate the review model and update the original specifications
Method	As outlined in [Daun et al. 2015c], the method can be briefly summarized as follows: 1. Validate the review model by performing perspective-based reviews and conduct negotiations with the stakeholders to reveal actual stakeholder intentions and to detect deficiencies in the review model. 2. Correct the review model in close collaboration between requirements engineers and function architects under consideration of the stakeholder intentions revealed. 3. Update the behavioral requirements and the functional design by applying automated model transformation techniques.
Input	Review model, stakeholders
Output	MSC specification
Condition/ limitation	-

Detect Defective Emergent Functional Behavior

Functional interaction

In function-centered engineering, existing functions from function libraries are reused, partly changed and updated, and combined with other existing or newly developed functions. The functional design is thereby built in such a way that the network of functions fulfills the systems requirements. In doing so, the functional behavior of the SUD results not only from the single functions but also from the interaction between the single functions.

Defective interaction

It is important to validate that all desired emergent behavior is provided by this functional interaction and that there is no undesired functional interaction. Therefore, a distinction must be made between real, undesired, defective functional interaction and functional interaction that was not specified in the requirements but can be considered as desired or at least as not undesired.

Main causes of functional interaction

Functional interaction may result due to several causes. For example, functional interaction can emerge from the following circumstances:

❑ *Overlapping control circuits* control, influence, or are impacted by the same context measurement. For example, a vehicle's speed is controlled by an adaptive cruise control system (as described in Chapter 2), the vehicle's speed will also influence the braking behavior of the electronic stability program, the automated locking and unlocking by the door control unit, and many more systems.

❑ *Multiple instances of the same system* commonly interact at runtime. For example, an avionics collision avoidance system of one airplane will interact with another airplane's collision avoidance system to avoid collisions between both airplanes, or automotive adaptive cruise control systems might be used to regulate autonomous traffic flows in traffic jam scenarios.

Explicit documentation

We suggest a solution concept to detect defective emergent behavior in a twofold approach. First, it is essential to explicitly document all relevant information. In the case of overlapping control circuits, this means that, for example, all effects and influences from functions towards context measurements and all sensing of context measurements by functions are properly documented within the functional design. In the case of multiple instances, it is important to specify the desired system behavior not only at the type level but also at the instance level. In doing so, we can define different configurations of the interacting instances (i.e., different numbers of involved instances or different parameterizations).

Second, based on the explicitly documented information, common graph analysis techniques can be applied to identify emergent behavior. In addition, model transformation techniques can be used to display emergent behavior in dedicated review diagrams (much akin to the review model presented in Section 8.3.2). More details regarding the automated analysis for overlapping control circuits can be found in [Daun et al. 2015b]; and for multiple instances in [Daun et al. 2015a].

Automated analysis

8.3.3 Validation and Analysis of Timing Behavior

Virtual Integration Test

During function-centered engineering, functions are defined which interact with other parts of the SUD or with parts of the system's context. These functions are typically decomposed across various layers of granularity and are related to other viewpoints using mapping and dependency relationships. For example, functions are mapped to logical components. These logical components may be refined from one layer of granularity to a lower one. As a consequence, the refined components will be related to implicit subfunctions which will have to be detailed and explicitly documented by the engineers. Furthermore, functions are also related to requirements. These requirements are often also refined when decomposing a function or a component, or when traversing from one viewpoint or granularity layer to the next.

Mapping and dependency relationships

For a valid system design, both kinds of relationships — between functions and requirements and between functions and components — require structural and behavioral compatibility. This means that the requirements of a decomposed function must be consistent and compatible with the requirements specified for its subfunctions. The same applies for transitions between viewpoints and granularity layers: for example, if a set of functions is mapped to a logical component, the requirements referring to the logical component have to be a valid refinement of the requirements referring to the mapped functions.

Valid system design for function relationships

The *virtual integration test* (VIT) [Gezgin et al. 2011] can be used to verify whether a composition or a transition between viewpoints or granularity layers is valid with respect to the specified requirements. To perform this VIT automatically, the specified requirements have to be formalized (i.e., translated into machine-

Validate relationships with a virtual integration test

readable specifications with well-defined semantics). With reference to Chapter 6, which addresses the problem of formalizing requirements in general, the building block in Tab. 8-4 describes how to derive formal contracts from requirements.

Tab. 8-4   *Method definition for the deriving formal contracts method*

SPES XT methodological building block	
ID	8.4
Name	Deriving formal contracts
Method	1. Separate requirements specifications into a) the assumptions under which the required behavior should be realized by the SUD, and b) the guarantees that the SUD shall give if all assumptions are met. 2. Make sure that the requirements do not specify any internal behavior of the SUD but only the black box behavior exposed at the ports of the SUD. 3. With the help of the methods from Chapter 11, derive formalized requirements from their textual representations. For formalized requirements to be checked automatically, it is necessary to define unambiguous semantics for each formalized representation of a requirement. Examples of such languages include the pattern-based requirements specification language defined in the CESAR project and its derivatives [Reinkemeier et al. 2011].
Input	Not yet formalized requirements specifications
Output	Semantically well-defined formalized contracts with separated assumptions and guarantees
Condition/limitation	Behavior is exposed and completely visible at the ports of the SUD

Challenges addressed   The method for deriving formal contracts directly serves as input for the multi-aspect virtual integration test. It also helps us to understand the context of the SUD and what exactly the SUD shall provide (in terms of multi-aspect behavior and non-functional properties). The VIT can be applied based on the formalized contracts. The purpose of the VIT is to check the validity of formal requirements that are related to functions in one of the following ways:

❑ *Compositional contracts*: contracts that are related to functions in a composition relationship — for example, a set of contracts A of a function in relation to another set of contracts B of its subfunctions

❑ *Contracts of different viewpoints*: contracts related to components of different viewpoints that are related to a function by trace links — for example, a set of contracts A of logical components and a set of contracts B of related technical components

❑   *Contracts of different granularity layers*: contracts that are re-
    lated to functions of different granularity layers by a trace link
    — for example, a set of contracts *A* of functions at system level
    and a set of contracts *B* of related functions at subsystem level

Internally, the VIT consists of two parts: the *compatibility check*   *Compatibility and*
and the *refinement check*. The *compatibility check* verifies whether   *refinement checking*
two given formal contracts are compatible, meaning that the func-
tions are composed without violating any of the contracts. This
becomes relevant if ports of the respective functions are connected
and the assumptions and guarantees refer to these connected ports.
The contracts and thus the functions are compatible only if the
guarantee of one function implies the assumption of the other function. The compatibility check can also be applied for more than two
contracts by checking their compatibility in pairs. The *refinement
check* verifies whether the composition of contracts of a set of
(sub)functions is a valid refinement of the contract of the parent
function. Both the compatibility and the refinement check ensure
that the integration of functions within a function hierarchy (or
between different granularity layers, or with components from another viewpoint) and the connection of functions with each other
with respect to the specified contracts are valid and consistent.

   The use of the VIT allows contracts addressing different aspects   *Timing and safety*
to also be considered [Gezgin et al. 2014]. Apart from functionality,   *requirements*
the VIT mainly concentrates on timing aspects. In addition, we can
also consider safety requirements as a second aspect. Timing and
safety requirements are often not independent and influence each
other. This influence is desribed in more detail in Chapter 11.

Timing Simulation and Analysis

The concept of timing paths enables a comprehensive and detailed   *Analyze response time*
analysis of the end-to-end system behavior. In early stages of the   *and data consistency*
engineering process, it is important to consider information regard-
ing estimated response times, data consistency, jitter, and robustness of the SUD. In particular, the evaluation of the requirements
linked to the timing paths is vital to ensure that a correct and safe
system emerges from the development process. Tab. 8-5 summarizes
the input and output of the method for analysis of the timing behavior.

Tab. 8-5   *Method description of the method for analyzing the timing behavior*

SPES XT methodological building block	
ID	8.5
Name	Timing path analysis
Method	Analysis of data flow behavior based on timing paths
Input	Timing path definition, functional network
Output	Analysis results (from simulation and schedulability analysis)
Condition/ limitation	The quality of the results depends on the completeness of the input

Consider interaction between process steps

Again, an example is shown for the adaptive light system (see Section 8.2.3): to ensure that the requirement to reduce the high beam within 0.5 second of the approach of another vehicle is satisfied, the method used has to consider the whole timing path between the camera recognition of the approaching vehicle and the adaption of the vehicle's light. Therefore, it is not only information on the single steps such as image processing, object recognition and detection, and the light adaption that is required, but, more importantly, the interaction between these steps.

Response time decomposition

A simple option for refining and verifying a timing path on a lower granularity layer is to decompose the required response time of the timing path to a lower level and assign each single step its own required response time. For example, each step receives its share of the global required worst-case end-to-end response time. If an analysis method ensures that each process step's specific share of the worst-case response time is always achieved, the end-to-end response time, which is the sum of these shares, will also be reached.

The problem of overestimation

Unfortunately, in most real cases, the distribution of allowed response times will lead to a significant overestimation. Hence, the use of simulation algorithms is an important means for timing analyses and prediction. Consider a simple system with a timing path having three consecutive tasks which share the same processor. Note that an actual processor does not exist at this stage of development, but the simulation technique can help at high level to determine whether the functional design is sound with regard to assumed hardware technology. Let us assume a high-priority load on the processor which occurs sporadically. Fig. 8-9 depicts a simulation by the tool *chronSIM* of the system showing three (consecutive) occurrences of the timing path. This simulation result allows us to identify that the sporadically occurring higher load will not result in global response time defects. The sum of the worst-case response times will include three times the higher priority load but it can only occur once in each instance of the timing paths.

Data consistency is another important aspect which can be eval-uated by analyzing the timing paths. During execution of a timing path, data can be lost, be processed multiple times, or be delayed due to excessive queuing. This may be due to scheduling, resulting jitter, different activation rates, or asynchronous communication. For example, in a timing path, data can be lost when the data ex-changed is realized by a shared variable. The value of the variable is overwritten by the subsequent execution of the same timing path before it is read by the subsequent processing step of the original timing path. This might result from a delay of the subsequent pro-cessing step or when two instances of one processing step are exe-cuted at almost the same time — for example, due to a previous delay. The same data can be processed twice in the same situation if the variable is accessed twice before it was updated by the subse-quent instance of the previous step. In particular, if there are several such consistency problems within the same timing path, only an appropriate simulation or analysis method will clarify which origi-nal data instances a result of the timing path is based on.

Consider data consistency

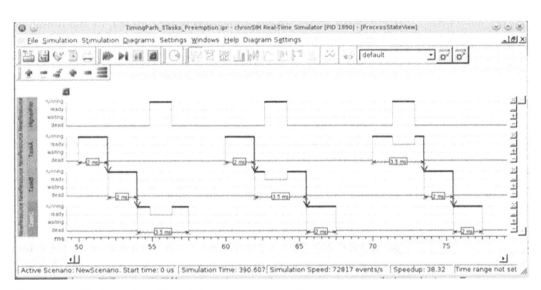

Fig. 8-9   *Timing path: problem of disturbing the response time*

The example requirement from the adaptive light system requires the light reduction within 500 ms when a vehicle is approaching. It would be incorrect to transfer this requirement to a requirement for

Investigate timing behavior early

the timing path of 500 ms. Instead, the system will have a frequency in which pictures of the camera are processed; this frequency will have to be part of the 0.5 seconds. So the 500 ms for the timing path will be reduced by the maximum distance between two camera pictures. If we allow 250 ms response time for the timing path, the frequency for the pictures also has to be 250 ms to ensure the overall deadline. If the timing path requires more time, the frequency for the pictures has to increase too. Hence, choosing the wrong frequency early in the design process results in a system overload. A lower frequency would have allowed for smaller processors. In conclusion, it is essential to investigate the timing behavior of the planned system as early as possible.

## 8.4	Summary

Challenges in function-centered engineering

Several challenges arise in function-centered engineering of embedded systems — in particular, due to the high degree of collaboration between different functions of one system or of multiple collaborating systems. For example, it is vital to the engineering of safety-critical systems to ensure all desired behavior will be executable, all undesired behavior is absent from the system, and that all requirements with respect to real-time aspects are fulfilled. In addition, it is desirable to support seamless integration between different engineering phases — for example, from requirements to functional design or from functional design to logical architecture.

Model-based documentation and analysis

We proposed a consistent and formal documentation format for the functional design as a basis for further development activities. As shown, the explicit model-based documentation is a promising approach on its own to support detection and correction of first obvious deficiencies. In a further step, automated techniques can be used to aim at solving detailed problem situations and to support seamless and consistent engineering of embedded systems.

Model-based documentation and analysis

The proposed techniques have been applied to industrial case studies and were discussed with industry experts to ensure industrial applicability and appropriateness (see [Daun et al. 2014c] for further details). In addition, experiments have shown that the proposed methods — such as validation by means of a review model (see Section 8.3.2) — can significantly increase effectiveness, efficiency, user confidence, and supportiveness (see [Daun et al. 2015d] for detailed results).

8.5   References

[Alfaro and Henzinger 2001] L. de Alfaro, T. Henzinger: Interface Automata. In: Proceedings of the 8th European Software Engineering Conference held jointly with 9th ACM SIGSOFT International Symposium on Foundations of Software Engineering (ESEC/FSE-9), ACM, New York, 2001, pp. 109–120.

[Büker 2013] M. Büker: An Automated Semantic-Based Approach for Creating Task Structures. In: PhD thesis, Carl von Ossietzky University of Oldenburg, 2013.

[Büker et al. 2011] M. Büker, W. Damm, G. Ehmen, I. Stierand: An Automated Semantic-Based Approach for Creating Tasks from Matlab Simulink Models. In: Proceedings of 16th International Workshop on Formal Methods for Industrial Critical Systems (FMICS), 2011.

[Büker et al. 2013] M. Büker, W. Damm, G. Ehmen, S. Henkler, D. Janssen, I. Stierand, E. Thaden: From Specification Models to Distributed Embedded Applications: A Holistic User-Guided Approach. In: SAE International Journal of Passenger Cars- Electronic and Electrical Systems, Vol. 6, 2013, pp. 194-212.

[Daun et al. 2014b] M. Daun, T. Weyer, K. Pohl: Validating the Functional Design of Embedded Systems against Stakeholder Intentions. In: Proceedings of the 2nd International Conference on Model-Driven Engineering and Software Development, SciTePress, 2014, pp. 333-339.

[Daun et al. 2014c] M. Daun, J. Höfflinger, T. Weyer: Function-Centered Engineering of Embedded Systems - Evaluating Industry Needs and Possible Solutions. In: L. Maciaszek (Eds.): Proceedings of the 9th International Conference on Evaluation of Novel Approaches to Software Engineering, SciTePress, 2014, pp. 226-234.

[Daun et al. 2015a] M. Daun, J. Brings, T. Bandyszak, P. Bohn, T. Weyer: Collaborating Multiple System Instances of Smart Cyber-Physical Systems: A Problem Situation, Solution Idea, and Remaining Research Challenges. In: Proceedings of ICSE WS - International Workshop on Software Engineering for Smart Cyber-Physical Systems (SEsCPS'15), 2015, pp. 48-51.

[Daun et al. 2015b] M. Daun, B. Tenbergen, J. Brings, T. Weyer: Documenting Assumptions about the Operational Context of Long-Living Collaborative Embedded Systems. In: W. Zimmermann, W. Böhm, C. Grelck, R. Heinrich, R. Jung, M. Konersmann, A. Schlaefer, E. Schmieders, S. Schupp, B. T. Widemann, T. Weyer (Eds.): Software Engineering Workshops 2015 (SE-WS 2015) - Gemeinsamer Tagungsband der Workshops der Tagung Software Engineering 2015, Vol. 1337, CEUR-WS.org, 2015, pp. 115-117.

[Daun et al. 2015c] M. Daun, T. Weyer, K. Pohl: Detecting and Correcting Outdated Requirements in Function-Centered Engineering of Embedded Systems. In: Proceedings of the 21st International Conference on Requirements Engineering: Foundations for Software Quality (REFSQ), Springer, 2015, pp. 65-80.

[Daun et al. 2015d] M. Daun, A. Salmon, T. Weyer, K. Pohl: The Impact of Students' Skills and Experiences on Empirical Results: A Controlled Experiment with Undergraduate and Graduate Students. In: Proceedings of the 19th International Conference on Evaluation and Assessment in Software Engineering (EASE), No. 29, 2015.

[Föcker et al. 2015] F. Föcker, F. Houdek, M. Daun, T. Weyer: Model-Based Engineering of an Automotive Adaptive Exterior Lighting System. ICB Research Report, No 64. University of Duisburg-Essen, 2015.

[Gezgin et al. 2011] T. Gezgin, R. Weber and M. Girod: A Refinement Checking Technique for Contract-Based Architecture Designs. In: Fourth International Workshop on Model-Based Architecting and Construction of Embedded Systems, 2011.

[Gezgin et al. 2014] T. Gezgin, R. Weber, M. Oertel: Multi-Aspect Virtual Integration Approach for Real-Time and Safety Properties. In: International Workshop on Design and Implementation of Formal Tools and Systems (DIFTS14), IEEE, 2014.

[ITU Z.120] International Telecommunication Union: Recommendation Z.120: Message Sequence Charts. International Standard, 2011.

[Reinkemeier et al. 2011] P. Reinkemeier, I. Stierand, P. Rehkop and S. Henkler: A pattern-based requirement specification language: Mapping automotive-specific timing requirements. In: Software Engineering 2011 Workshopband, Gesellschaft für Informatik e.V. (GI), 2011, pp. 99-108.

[Vogelsang et al. 2012] A. Vogelsang, S. Eder, M. Feilkas, D. Ratiu: Functional Viewpoint. In: K. Pohl, H. Hönninger, R. Achatz, M. Broy (Eds.): Model-Based Engineering of Embedded Systems: The SPES 2020 Methodology. Springer, Berlin Heidelberg, 2012, pp. 69-84.

Karsten Albers
Matthias Büker
Laurent Dieudonné
Robert Hilbrich
Georgeta Igna
Stefan Kugele
Thomas Kuhn
Maike Rosinger
Raphael Weber

9

Optimal Deployment

The SPES XT modeling framework combines multiple modeling viewpoints and offers an effective means for modeling and exploring various embedded systems designs. One crucial design step is to deploy the logical architecture to the technical representation of the physical system constituting the embedded system architecture. Since various system requirements and design goals impact the development, we describe several methods and techniques for mapping, analyzing, and validating different deployment solutions. Finding an optimal deployment is a difficult process due to a large number of often conflicting requirements that should be satisfied. The requirements address resource, timing, safety, deployment, economical, and regulatory aspects. This chapter presents design space exploration methods, including their application to an automotive example.

© Springer International Publishing AG 2016   145
K. Pohl et al. (eds.), *Advanced Model-Based Engineering of Embedded Systems*,
DOI 10.1007/978-3-319-48003-9_9

9.1   Introduction

Deployment The quality of today's embedded systems — for example, in vehicles, airplanes, or automation plants — is heavily influenced by their overall architecture. This architecture subsumes the logical and technical structures of an embedded system as well as the mapping of the logical to the technical structure, which we call deployment. In the scope of this chapter we address challenges such as: On which target execution unit do we execute *which* software component as a task? *Which* sensor/actuator shall be connected to *which* distributed I/O unit? Finding a feasible deployment (meeting all constraints) may be difficult; designing an optimal embedded system is an even bigger challenge.

Optimality Deployment is merely the mapping of a logical architecture to a hardware architecture that satisfies all constraints (e.g., resources, timing, safety, etc.). *Optimal* deployment considers the minimization/maximization of certain objectives such as costs, weight, etc. In order to solve the optimal deployment problem adequately, we have to define a notion of optimality (i.e., optimization objectives). In practice, there are usually many objectives, some of which are conflicting.

Design space exploration Typically, during system development — regardless of the application domain — engineers face many potentially conflicting requirements. For example: the goal to develop an airplane with a better ratio of kerosene consumed per aircraft passenger is more expensive due to modern materials or jet-propulsion engines. The set of *possible* solutions is called the design space, whereas the set of feasible solutions is called the solution space. The goal of *design space exploration* (DSE) is to find the best compromise (trade-off) amongst the different goals (optimization objectives) in the solution space [Eisenring et al. 2000]. This results in a multi-objective optimization problem [Hamann et al. 2006].

Optimization methods Traditionally, this problem was approached with a lot of experience and gut feeling [Diebold et al. 2014]. Of course, on the one hand, the influence of engineering experience should not be underestimated; on the other hand, however, these solutions are — if at all — optimal by accident. Experience has taught us that it is necessary to recheck all constraints (e.g., robustness, safety, maintainability, and so on) whenever we change a design. Therefore, methods leading to an optimal architectural design are desirable.

9.1.1   Challenges regarding Optimal Deployment

For an optimal deployment, the main prerequisite tasks include capturing and formalizing the following: the needs of the system functions to be deployed, the capabilities of the execution units that can potentially be used for the targeted architecture, and the optimization objectives which help to decide which design is *better*. These properties drive the deployment problem. They have to be expressed in a way that allows them to be compared, and they should be capable of being processed by machines in order to automate some design decisions.

Prerequisites

Another challenging aspect is to understand, capture, and again formalize the ideas and principles employed by the experienced system engineers to design valid system architectures. In particular, the rationale behind certain design decisions should be extracted and converted into computationally solvable problems. For example: use the same hardware architecture as a blueprint for the next generation.

Understanding engineering experience

Trade-offs are necessary to decide which solution is better in a specific context. Optimization objectives are measured with a metric and are usually expressed with phrases such as "the more, the better", or the opposite "the less, the better", for example, when considering costs. A dedicated threshold is not essential for an objective, but often there is one — for example, a maximum weight that must not be exceeded but the further below that weight "budget", the better. As stated above, the objectives often conflict, which means that improving one objective worsens another (e.g., weight and costs).

Trade-offs and budgets

In contrast to optimization objectives, constraints describe properties that have to be fulfilled. Here it is sufficient to just reach the required constraint — there is no need to do better than that. Constraints can also be optimization objectives and act as a threshold that marks the border from an invalid solution to a valid one and towards an even better one. Other constraints can be given to define possible target architectures consisting of execution resources that may be consumed if some functionality is deployed to them. There are also mapping constraints derived from requirements such as segregation, dissimilarity, or other safety properties.

Deployment constraints

All of these properties, along with a sound understanding of how they influence each other, constitute the design space. The main challenge addressed in this chapter is how to explore this vast design space in an efficient and effective way to find feasible and

Challenge: exploring the design space

optimized solutions. This results in two DSE challenges which we address in this chapter:

1. Find an optimal design
2. Find a feasible design

No ideal solution   Since the problem of finding feasible and optimal designs itself is too difficult to solve in an ideal way, there are several approaches which support the designer in finding an optimized deployment as well as a feasible design.

9.1.2   Methodological Framework

Generic DSE framework   To address the prerequisite and efficiency challenges from Section 9.1.1, and in order to offer a common interface for concrete DSE techniques, here we integrate and apply the design space exploration approach in the typical development process for software-intensive embedded systems. We define a *generic DSE framework* (see Fig. 9-1) related to the SPES XT modeling framework described in Chapter 3 and Chapters 6, 8, and 10.

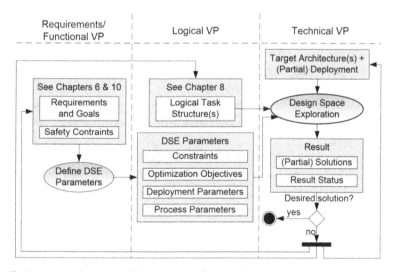

Fig. 9-1   *Overview of the generic DSE framework*

DSE context   The goal of this framework is to allow the integration of a set of specific DSE techniques (presented in Section 9.3) by offering a common interface to address the two main DSE challenges (find an optimized solution and find a feasible solution). Fig. 9-1 also gives an overview of the relationships to other challenges.

The requirements viewpoint specifies goals, requirements, and constraints which need to be refined into DSE constraints, optimization objectives, deployment parameters, and process parameters during the design phase if they are relevant for the DSE process. This usually also includes a more concrete specification and formalization of requirements such that the requirements can be processed by an automatic DSE method. For example, a safety requirement might lead to a DSE constraint stating that certain tasks are not allowed to be allocated to the same computing resource to avoid a single point of failure (dislocality). In addition, DSE constraints may be derived from requirements that address different aspects such as timing, safety, security, power, and also more abstract goals such as reusability, verifiability, and testability. For the latter in particular, it is important to note that these goals have to be specified formally to enable direct support by an automatic DSE method. For example, this may be achieved by defining a respective metric. Typical goals such as "minimize system costs", which refer to one or several concrete optimization functions, utilize cost metrics whose values shall be minimized as, for example, costs of electronic control units and costs of cables.

Relationship to the requirements viewpoint

The modular safety assessment (see Chapter 10) may contribute to the process of defining DSE parameters by deriving safety constraints from requirements based on the results of safety analysis techniques such as *fault tree analysis* (FTA) or *failure mode and effects analysis* (FMEA). The early validation of engineering artifacts (see Chapter 6) may also play a role in this process step — for example, by offering techniques for formalizing textual requirements, which is an important step in deriving constraints needed by automatic DSE methods.

Contributing methods

Process parameters may be defined in order to control the automated parts of the DSE process. These parameters may also be derived from a high-level requirement or be defined in this process step. An example of a process parameter is a timeout for the DSE process which restricts the time spent searching for an optimal solution to a maximum of 5 hours after which the optimization process shall be aborted.

Process parameters

The process step *Define DSE Parameters* covers all steps needed to decide which requirements influence the deployment process. This also includes refining the affected requirements to a sufficient level of detail to be able to apply the DSE process. This may be done either manually by the user or supported by tools, such as wizards. The consideration of user support methods is not covered

User support

in this chapter but there are some works that address this topic — for example, [Rosinger et al. 2015] and [Rosinger et al. 2014].

Logical components and their communication

In order to apply the DSE methodology, a logical architecture is necessary to identify the deployable entities in terms of logical tasks and signals (logical task structure). In addition, the communication (signals between tasks) has to be modeled explicitly for the DSE methods because those signals have to be allocated to communication resources such as buses if the respective tasks are not allocated to the same computing resource. In addition to the task structure, a set of non-functional properties also has to be defined — for example, concerning the activation behavior and resource demand of tasks and signals. This is important for the second DSE challenge of finding a feasible solution.

Task structure

Generally, the logical task structure is derived from a functional model. In this transition step from the functional to the logical viewpoint, the designer has to decide which functions should be bundled into one logical component and which are the atomic logical components (tasks). See Chapter 8 for more details.

User-guided process iterations

The execution of a single DSE process run does not necessarily lead to a valid and complete solution that is also satisfactory for the user. In practice, therefore, such a process will typically be repeated several times until a solution is found that fits the user's needs. In Fig. 9-1, we outlined some typical process iterations that detail how the DSE process may be repeated. Another important step is the evaluation of the results of one process iteration to assist the user in assessing the proposed solutions and deciding how to continue the process. Again, there are methods and wizards to support the user in this step (see [Weber et al. 2014] and [Rosinger et al. 2015]), although this issue is not addressed in this chapter.

Finishing and restarting the DSE process

Based on the evaluation results, the user may choose one of the proposed solutions best fitting his needs. If there is such a solution, the process is finished and the chosen deployment is returned as the result. If there is no solution, the user may modify the technical target architecture(s) including the deployment (e.g., based on the intermediate solutions and evaluation results), adjust the requirements and goals, or directly modify the DSE parameters and then start the DSE process again.

Refinement of input artifacts and DSE parameters

There are several possible intervention points. For example, the user may decide (based on an intermediate solution of the previous DSE process run) that the specification of goals was not complete and that another goal needs to be added — for example, a requirement to minimize the cable length of the DSE system. As a next step, the user may also add or refine constraints and process parameters or adjust the optimization objective. Here, for example, the concrete characteristic of the optimization function may be adjusted or the prioritization or rating of different optimization objectives may be changed. In addition, the task structure and target architecture may be modified. For example, the allocation of a certain subset of tasks that have been successfully deployed in a previous solution may serve as the initial deployment for the next process iteration. Only the remaining tasks not yet deployed are then considered by the DSE.

The SPES XT tool platform (see Chapter 14) comes into play once the DSE methods described in this framework have been implemented and the subjects of tool interoperability and exchange of model artifacts become relevant. The generic DSE framework already defines an interface for input and output artifacts of DSE methods, which is put in more specific terms in the method descriptions of Section 9.3 by using a common template. These templates provide the basis for defining specific service descriptions for the SPES XT tool platform.

Relationship to the SPES XT tool platform

9.2   Extensions to the SPES Modeling Framework

This chapter details how the SPES modeling framework has been extended to foster optimal deployment in the SPES XT modeling framework. In the center of Fig. 9-1, the abstract interface in terms of input and output artifacts of the generic DSE process is depicted denoted as *DSE Parameters*, *Logical Task Structure(s)* and the technical *Target Architecture(s),* possibly including a *(Partial) Deployment.* The output of this process is a result that contains a set of *(Partial) Solutions* and a *Result Status.* In the following, we will explain the artifacts and describe their role in the process of finding an optimal solution.

Abstract interface of input and output artifacts

A mandatory input for the DSE is a logical task structure consisting of communicating atomic components denoted as tasks that are to be executed on the available hardware resources. In general, the task structure is derived from a functional model by deciding which functions should be bundled into one task. This task structure also contains information about the communication between each task — how often communication takes place and how much information is exchanged. The logical task structure is completely reflected by the logical viewpoint of the SPES modeling framework.

Logical task structure

Target architecture   Another mandatory input artifact for the DSE is a technical hardware architecture (denoted as the target architecture) which models the resources to which tasks should be deployed. A target architecture may contain a (partial) deployment relationship referring to tasks already deployed. This is because experience shows that in some domains, embedded systems are often developed incrementally. This means that systems are built on top of an existing model. In order to reuse such existing models, a target architecture may consist not only of a hardware architecture which offers certain services and interfaces, but also of tasks already deployed which represent the functionality of the existing design. The technical hardware architecture can also be completely represented with existing SPES modeling framework artifacts.

DSE parameters   The remaining inputs are summarized as *DSE parameters*. They represent constraints, optimization objectives, deployment parameters, and process parameters; there may also be additional classes of parameters not explicitly listed here. *Constraints* specify which DSE system configurations are considered to be valid, thus restricting the design space. *Optimization objectives* define the attributes that should be considered during the DSE to rate solutions in order to find optimal deployments. In contrast, *deployment parameters* do not define what should be optimized but rather describe the parameters that may be set or changed during the optimization process and thus the degree of freedom for the DSE method. For most DSE techniques, this is primarily the deployment itself. We also define *process parameters* which capture all parameters that influence and control the automated parts of the DSE process itself without relating to the system under development.

DSE results   The output of the DSE process consists of a set of valid (but potentially incomplete) solutions and a result status. Solutions are valid (correct) if and only if all given constraints are satisfied. Completeness means that all tasks of the given task structure are allocated to the technical architecture. If there are no valid (partial or complete) solutions at all, the DSE may either return an empty solution set or the unmodified technical architecture as specified at the start of the previous DSE step. An incomplete solution may serve as an intermediate result and the DSE process may iterate with adjusted inputs and parameters to obtain a better and (possibly) complete result. If valid (partial or complete) solutions exist, the DSE returns the best solutions for the optimal deployment problem w.r.t. the defined optimization objectives.

Additional DSE artifacts   In addition to the artifacts visible at the interfaces of the general

DSE framework, there are some intermediate artifacts which relate constraints and optimization objectives to the corresponding architectures (logical and technical). In relation to the SPES modeling framework, constraints and optimization objectives are a specialization of requirements with an explicit relationship to metrics which in turn refer to parameters. Fig. 9-2 gives an overview of the additional DSE artifacts and their relationships to SPES modeling framework artifacts.

Metrics

A metric defines a function which combines multiple properties of a system (referred to as the relationship to abstract parameters) into a characterization. An example would be the combination of multiple cost values of subcomponents, where the sum of all costs denotes the overall system costs. Other metrics may be more complex. Extensions may therefore be defined to denote more complex metrics.

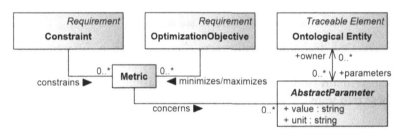

Fig. 9-2   *DSE-specific SPES modeling framework extensions*

An abstract parameter is supposed to be a superset of all parameter artifacts. It defines the relationship to ontological entities but leaves the actual property or parameter definition undefined. The *value* and *unit* properties, however, must be present so that a metric may directly refer to them. Abstract parameters may be refined into system parameters referring to system properties or measures. An abstract parameter may also represent a process parameter which constrains the scope or runtime of a design space exploration process as mentioned above. With the abstract parameter and the metric artifacts, the transitive relationship between constraints and optimization objectives and ontological entities (which may represent the logical or technical architecture) is complete. The following section will detail methodological building blocks for addressing the optimal deployment challenges.

9.3   Methodological Process Building Blocks

In the following, the two main DSE methods and suitable backend techniques are described. These methods and techniques constitute the building blocks for our DSE methodology ("method toolbox").

Two DSE steps

To understand the techniques and how they contribute to the optimization of deployments, it is important to explicitly define the two steps which directly correspond to the two DSE challenges above. There are different ways to approach this. Usually, a DSE method tries to find a "good" starting point that satisfies all constraints (is feasible) and then optimizes it in subsequent steps. Some DSE methods do not try to find this starting point and instead require the engineer to provide it. In either case, due to the non-linear shape of the solution space within the design space, it is impossible to assess how close to the optimum these starting points are. Therefore, in general, there are two steps:

1. Synthesize optimized solution
2. Find feasible solution/validate solution

Different techniques

Every DSE technique implements these steps in a different way. Because each technique is required to be very efficient (the goal is to test as many solutions as possible for feasibility and optimality), these two steps are closely integrated. Fig. 9-3 gives an outline.

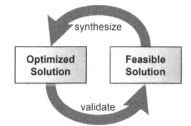

Fig. 9-3   *Two typical iterative DSE steps*

Diverse constraints and optimization objectives

The set of possible constraints and optimization objectives along with the metrics are very diverse. Hence, each technique performs differently depending on the complexity of these steps. For example: synthesizing a feasible solution for a priority-based scheduling policy may be more difficult than for a time slot-based scheduling policy. In the following subsections we describe the above-mentioned two steps in more detail along with the techniques that address them.

9.3.1   Finding an Optimized Spatial Deployment

The goal of finding an optimized spatial deployment is to explore the design space to find an optimized and feasible (to some degree) solution. The feasibility explicitly excludes timing consideration since checking the feasibility of timing parameters requires a schedulability analysis which itself is too complex to be part of this method. At most, we can cover abstract time resource budgets to guide the DSE. This is also why the optimization may have to abstract certain objectives in order to find a solution.

Optimization without schedulability

As stated above, the DSE is a very difficult challenge and it is so complex that for each additional optimization objective, the size of the design space grows exponentially. This is why we first look for a spatial deployment and then find out whether the solution is completely feasible. A spatial deployment denotes a mapping of tasks to a target architecture where each required property of a task is satisfied by a provided property. For example, if a task T_1 requires a direct input from a sensor S, all execution units with a sensor of type S are valid deployment targets for T_1. Another example: if another task T_2 requires a certain execution unit-specific extension E, it may only be deployed to execution units with an extension of type E. Depending on the complexity of these constraints, it may be easy to find an exact solution (optimal with respect to the accuracy of input properties). However, if the constraints generate a rather large solution space, it may be a good idea to use heuristics which only deliver some parts of the solution space. Ideally, these are only parts for which we know (from engineering experience) that they contain mostly optimized solutions.

Optimized mapping respecting simple constraints

9.3.2   Finding a Feasible Schedule

After finding a deployment that meets optimization objectives and simple constraints, the correct timing behavior has to be guaranteed. There are a number of techniques to ensure a feasible timing behavior. This is important in particular for safety-critical systems that only work properly if the right information is provided at the right time. Control algorithms are an example of periodically updated sensor data which has to be processed within a certain time limit in order to provide actuators with the necessary data to control environment properties. Since these systems rely on some feedback from their surroundings, this results in distinct timing requirements for the processing part of the chain. Finding a feasible schedule which meets these timing requirements is thus very im-

Validating deployment-dependent constraints

portant. In some cases, proof that these time limits are met is required. Depending on the degree of distribution of the whole system and on the scheduling policy, it may be difficult to integrate the complete schedulability analysis into a deployment optimization method. Hence, we distinguish between finding an *optimized spatial deployment* and a *feasible schedule*.

Scheduling strategies

In the context of a feasibility check there are different classes of techniques that cover different scheduling strategies:

a)   Utilization-based (e.g., estimation)
b)   Static scheduling (e.g., timetable-driven)
c)   Dynamic scheduling (e.g., fixed priority)

Comparison of strategies

Since techniques of class a) perform a more abstract and approximate estimation of timing properties, their preconditions are typically not as strict as the preconditions of techniques from classes b) and c). Thus, it is possible to perform these techniques before applying techniques from classes b) and c). In those cases, the techniques from classes b) and c) may benefit from the previous application of a technique from class a) because certain design decisions were already taken, resulting, for example, in additional constraints. The techniques from classes b) and c) cannot typically be combined with each other consecutively because they address different scheduling policies. In addition to the consecutive combination of techniques, where the compatibility of interfaces has to be considered, it is also possible to repeat complete technique sequences. However, this only makes sense if new information was derived from the previous iteration. There are a number of techniques that implement these methods; they are explained in the following.

9.3.3   DSE Techniques

Automatic Deployment

Purpose/goals

The automatic deployment technique is based on an *integer linear program* (ILP) optimization approach. The problem is transformed into a representation as a system of mixed integer linear equations. This equation system is solved by an ILP solver taking defined optimization objectives into account.

Preconditions

The ILP optimization approach requires a conforming input model that defines logical tasks and communication dependencies. Furthermore, the target architecture consists of runtime platforms and communication links. To retain flexibility and enable rapid adaptation to new network types, link capacities are specified using an additive model. Additional project-specific resources such as memory and CPU capacities may be specified for target platforms.

Explicit constraint specifications permit developers to control the output of the optimization algorithm. Currently, basic Boolean constraints (and/or/not/implications) as well as set-based constraints (forAll and atLeastOne) are supported. Recursive use of these constraints enables the definition of more relevant deployment constraints that limit the solution space for the ILP algorithm. Examples of constraints include the following:

Supported constraints and optimization objectives

❑   Function A must be deployed to computer X
❑   Function B must be deployed to a computer with attribute Y=Z
❑   Functions A and B must be deployed to different types of computers

Constraint definitions are usually provided by system designers as part of ILP definition files. Optimization criteria request that particular variables take maximum or minimum values. By default, only one variable is optimized. If multiple variables are to be optimized at the same time, a target variable that combines those variables using constant weight factors has to be provided by developers.

The ILP optimization algorithms return a deployment matrix that defines whether a function is deployed to a particular target environment or not. This technique also states whether a physical entity is used or not.

Outputs

The mapping of a deployment problem to an ILP description is supported by a generic ontology that captures all relevant properties of the input models. It serves as the target for a tailored model-to-model transformation, for example, from domain-specific languages.

Method description

The transformation of conforming model elements into an ILP format enables the definition of optimization criteria and constraints using the native language of the CPlex Optimizer[3]. In this way, basic domain knowledge that is invariant to deployment projects is provided. Furthermore, it is possible to add tailored transformations from the input model that generate further, project-

---

[3] http://www.ibm.com/software/integration/optimization/cplex-optimizer/

specific constraints that enable detailed control of the deployment process. The overall process of this technique is depicted in Fig. 9-4.

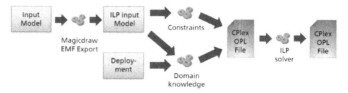

Fig. 9-4   *ILP-based deployment process*

Tool support   Currently, the ILP-based deployment algorithm is realized as an Eclipse-based tool that transforms input models into ILP description files which are read by the commercially available CPlex Optimizer.

Deployment under Consideration of Timing Optimization

Purpose/goals   The optimization of the timing behavior is important for a robust and reactive system. Even with a fixed deployment, parameters such as priorities, execution orders, or offsets can require an optimization to keep the necessary timing constraints. For the optimization, meaningful feasibility evaluation methods are required.

Preconditions   A precondition for optimizing the scheduling parameters is a fixed deployment, otherwise the logical tasks and the planned architecture are required. The task model can, for example, consist of AUTOSAR software components with their executable entities which should be distributed to an architecture consisting of single or multi-core processors.

Supported constraints and optimization objectives   Timing constraints such as end-to-end worst-case response times (for timing paths), reaction times, local response times, utilization limits, constraints on the execution rate, or the data consistency (e.g., exclusion of data loss) and more are supported. Optimization objectives can be, for example, the reduction of the average response time or a balanced utilization of resources.

Outputs   An output of the optimization is one or more complete models, including the deployment and scheduling parameters which were selected. In addition to the model, detailed simulation and analysis results of the behavior of the optimized model are available, including the evaluation of the timing constraints.

Postconditions   The postconditions are that the resulting model is consistent and can be simulated or analyzed and that the timing requirements are satisfied (depending on the concrete problem).

Method description   The timing optimization consists of a set of extendable optimization methods supported by methods for the timing simulation and schedulability analysis. The scheduling parameters must be optimized to achieve meaningful and precise information on the quality of a candidate system in terms of the timing behavior. The real achievable response times and other timing objectives can only be calculated for systems with optimized scheduling.

The proposed timing optimization can use round-based metaheuristic optimization approaches such as simulated annealing [Kirkpatrick et al. 1983], tabu search, and others. To allow fast and reliable execution, the optimization uses the detailed evaluation results from the simulation and schedulability analysis of one candidate solution to suggest further optimizations, promising modifications, or candidate solutions. The optimization is guided by detailed evaluation results. *Timing optimization*

One possible optimization method is the distribution of functionality to a target architecture (deployment). Several very different problems can be summarized under this method depending on the specific models and goals — for example, the distribution of AUTOSAR software components or the static distribution of tasks on a multi-core processor.

Candidate solutions based on a fixed deployment still have several degrees of freedom in terms of scheduling parameters such as priorities, offsets, orders of executable entities, etc. Good values are required here to allow a meaningful evaluation of the candidate solutions, a realistic statement on the fulfillment of timing requirements, and to realize a reliable and robust system in terms of timing.

For the timing optimization technique described in this section, an implementation framework is available which allows quick adaptation of the optimization process to specific challenges and the addition, for example, of other constraints. It is realized in the tools chronOPT (see *http://www.inchron.com/tool-suite/tool-suite.html*) from INCHRON which utilizes chronSIM (for simulation) and chronVAL (for schedulability analysis) for evaluation purposes. The goal of this tool suite is the development, investigation, and optimization of the timing behavior of embedded systems on several levels of the design process. *Tool support*

Deployment Based on Correctness by Construction

This engineering technique is intended to support system and/or software architects in achieving spatial and temporal deployments. The method is based on the concept of achieving correctness by applying formalized approaches for the construction of artifacts *Purpose/goals* using *constraint programming* [Chapman 2006], [Hall and Chapman 2002].

Preconditions   Constraint programming requires a precise and formal problem specification. The input distinguishes between logical and technical components in order to separate the problem specification from the solution description. Software applications are instances of logical components and hardware resources are structured into technical components. The function network is expressed as a logical task structure with a representation of the requirements of each software application (hardware resources, safety, timing, etc.). Each software application consists of at least one thread. All threads of the same software application have identical hardware requirements. Relationships (e.g., dislocality) always apply to all parts in a parallelized software application. A hardware description must adhere to a tree topology, where the cores are the leaves of the tree; the processors are one level higher; above these are the boards, etc.

Supported constraints
and optimization
objectives   The optimization objectives focused on in this technique are the number of execution units and an optimal scheduling of the software application's threads on the selected hardware resources. The technique deals with quantitative and qualitative constraints, such as capacity (memory and CPU), safety and criticality level, dissimilarity, dislocality, communication proximity, I/O type, and timing requirements (period, jitter, delay, timing synchronization).

Outputs   The technique produces different valid spatial and temporal deployment solutions. A valid spatial deployment is represented by a mapping of software applications to hardware resources (on cores, but also as occupation and/or consumption of I/Os, memory, communication and bus bandwidth, etc.). A temporal deployment is a scheduling of each application thread according to the timing and performance requirements. It includes (amongst other things) time slices, computation of a hyper-period, definition of tolerable jitter, additional delays for initialization tasks, execution, etc. The different spatial deployment variants are submitted to the system/software architect who compares and sorts them by applying different evaluation criteria based on metrics. Several metrics are predefined, such as *uniform cCore load distribution*, *max. freecore capacity*, etc., and new ones can be also defined by the architects. The temporal deployment is calculated only on the chosen spatial deployments.

Method description   In order to facilitate and simplify the input of a mapping specification for the user, textual domain-specific languages were developed. The grammars of these languages describe the correct syntax for mapping and scheduling specifications. These languages constitute a "bridge" between the domain-specific context of a mapping/scheduling challenge and the technical context of a constraint solver. The specification for a mapping/scheduling problem is subsequently transformed from the internal data model into a constraint satisfaction problem (CSP). If there are solutions for the specified CSP, these solutions will be used to enrich the internal data model with allocations of software components and utilizations of hardware resources, such as processing cores, memories, or IO adapters. More details are presented in [Hilbrich and Dieudonné 2013].

The tool PRECISION PRO was developed by Fraunhofer FOKUS to evaluate and illustrate the *correctness by construction* technique. It was developed on an Eclipse 4 Rich Client Platform. It is based on the Eclipse Modeling Framework (EMF) and the OSGi Implementation Equinox. The constraint solver *firstCS* from Fraunhofer as a form of a Java library has been used to perform constraint programming [Hofstedt and Wolf 2007], [Wolf 2006].

Tool support

Deployment and Scheduling Synthesis Using SMT Solving

Purpose/goals

The search for solutions within the DSE parameter space can be automated using highly efficient satisfiability solvers. In this section, we present a DSE technique based on *satisfiability modulo theories* (SMT) [Barret et al. 2009] which, depending on the DSE constraints, can be used to synthesize valid schedules for a given deployment or valid deployments including schedules for a given target architecture, or to identify a suitable platform and the corresponding deployments and schedules.

Preconditions

The input for this technique consists of a set of tasks, a target architecture, a set of optimization goals, and, optionally, a partial deployment. The set of tasks is annotated with precedence constraints and further attributes such as their individual resource claims or safety classifications. The technical architecture consists of a topology of computing units (e.g., cores of a multi-core platform) connected by communication and storage resources (e.g., buses, shared memory, etc.). Design parameters can also be included — for example, safety levels or energy constraints. In addition, the designer may add constraints obtained from previous system designs regarding permitted or forbidden mappings of tasks onto computing units. The method uses a time-driven approach — that is, the user should specify timing properties that represent the worst-case execution time (WCET) for each task. A WCET of a task represents the worst/longest time in which the task should be executed. Each computing unit can process only one task at a time (single core). Moreover, a task cannot be interrupted during execution, which means that task pre-emption is not allowed. Communication between tasks deployed on separate computing resources is ensured by signals sent via a shared bus. The user can also specify a worst-case transmission time for each signal.

Supported constraints and optimization objectives

The technique supports different optimization objectives for increasing overall system performance in terms of time and resource usage (e.g., minimize the end-to-end latency, minimize the number of computing resources). In general, this enables a Pareto-optimal solution of the system design.

Outputs

The technique generates valid and complete deployments which respect the precedence order between tasks and fulfill all timing and resource constraints. Each deployment includes a set of feasible schedules. Furthermore, optimal solutions are obtained when the search terminates within the time limits defined by the user. Each schedule provided for a deployment is unique for the allocation; if two schedules have equal end-to-end latencies, there is at least one task or signal that is scheduled differently in the two schedules.

Method description

The inputs of the deployment problem are formalized as logical and arithmetic formulas encoded using the SMT-LIB2 language [Barret et al. 2010]. An SMT solver is then used to synthesize valid and complete solutions. To find a deployment based on SMT solving, we generate a set of formulas over variables that represent the start time and end time of tasks and signals. The precedence order between tasks and signals is transformed into constraints between these variables. In addition, we specify constraints between these variables to prevent time overlaps for tasks allocated on the same resources. Furthermore, we also generate formulas over variables representing the mapping of each task or signal to the platform resources. If the mapping is known at the beginning of the deployment process, the corresponding variable is bound to a fixed value, ensuring that any valid solution will preserve it. Otherwise, the SMT solver chooses a valid value for this variable.

DSE evaluation options

To find one solution, the SMT solver is called only once. However, to find optimal solutions, we apply either a binary search algorithm or a guided-search algorithm. The latter algorithm requires that, during each round, the SMT solver is queried for a shorter schedule than the one obtained in the previous round.

With this technique, the DSE phase can result in refining the deployments between tasks and the computing units, or in adjusting the target architecture or the task structure. The overall process is implemented in an architecture wizard enabling provision of a step-by-step refinement of the system design.

AutoFOCUS3 [Kondeva et al. 2013] implements this method. *Tool support* Moreover, each viewpoint of the SPES XT modeling framework is represented as a separate module in the tool. AutoFOCUS3 is implemented on the Eclipse Rich Client Platform and employs Z3 [de Moura and Bjørner 2008] for SMT solving.

Two-Tier Iterative Design Space Exploration

This technique finds valid and near-optimal deployments of a set of *Purpose/goals* currently undeployed software tasks onto a given base target architecture to which other tasks might have already been allocated.

As input, this technique requires a target architecture that con- *Preconditions* sists of a global TDMA scheduled backbone bus (e.g., FlexRay) which connects several clusters, each of which contain processing units which in turn are connected by a local priority-based bus (e.g., CAN). Each cluster thereby contains one processing unit which acts as a gateway to other clusters. The rest of the input artifacts are the same as defined in the general DSE framework.

Supported constraints include timing constraints (end-to-end and *Supported constraints* local deadlines) and deployment constraints. Supported optimiza- *and optimization* tion objectives include monetary hardware costs, power consump- *objectives* tion, the number of processing units used, the weight of hardware resources, and the volume/size of hardware resources.

One output is a model which consists of a hardware architecture *Outputs* resulting from applying a (possibly empty) set of modifications of the set of allowed modifications to the base hardware architecture. Furthermore, the model consists of the task structure that was input into the technique and a (partial or total) deployment relationship. Another output artifact of this process is a result status detailing the status of the DSE process, that is, whether a timeout occurred, if the solution is partial or complete, if an error occurred, etc.

The deployment relationship is valid and is a refinement of the *Postconditions* input deployment relationship, meaning that the allocation of tasks that were already deployed in the existing system is identical. Furthermore, the system is consistent and can be scheduled.

To cope with the complexity, the DSE process is divided into *Method description* two steps (global/local analysis). The first optimizes the system level and the second optimizes each hardware subsystem separately.

The goal of the global analysis is to distribute the set of unallo- *Global analysis* cated tasks among the different subsystems of the hardware architecture of the existing system. The computational capacity of each subsystem is calculated by aggregating the computational capacities of its processing units. On the global analysis tier, as a notion for computational capacity we use the maximum fraction (share) of processor time a task would need from the whole processor time (it is assumed that each processor can provide 100% computational capacity) when running on a processor. Based on this metric, the costs needed to place a set of tasks on a subsystem is estimated. The resulting global pre-allocation serves as input for the local analysis.

Local analysis   The local analysis is performed separately for each hardware subsystem. It allocates the tasks that were mapped to the subsystem to the processing units of that subsystem but without exceeding the subsystem's hardware cost limit that was predicted by the global analysis. In this step, an exact characteristic of computation capacity is used which takes into account that each allocation has to satisfy a full-blown schedulability analysis to be considered feasible. If there are tasks that could not be deployed to a subsystem, they remain undeployed and a so-called "backtracking" phase is started. The result of the local analysis step is a local deployment of tasks to processing units, which may be incomplete, denoted as partial deployment.

Backtracking   In the backtracking step, tasks that could not be deployed to processing units within the proposed subsystem under the given cost limit are distributed to the hardware subsystems again by re-applying the global analysis. Here, only the undeployed tasks are considered, while the allocation of already deployed tasks remains unchanged.

Tool support   There is an Eclipse/Papyrus-based UML/SysML implementation for the specification of the DSE-specific modeling elements and it provides an input for the DSE backend tool called Zerg [Thaden 2013]. Zerg itself translates the global and local analysis parts into different backend formats (e.g., an MILP backend) which are solved by a respective solver (e.g., Gurobi or CPlex). Depending on the results and the configuration of Zerg, the backtracking step triggers another DSE execution. More details on the global/local analysis and the backtracking can be found in [Büker et al. 2011] and [Büker et al. 2013].

Flexible Deployment Respecting User Objectives and Constraints

Purpose/goals   The goal of this approach is to give the user of the deployment tool as much freedom as possible in terms of *what* to optimize and *which* constraints to respect. Therefore, rather than specifying how to obtain an *optimal* solution, the approach specifies *what* it should look like. This is achieved by providing a flexible solver infrastructure in combination with the domain-specific language *SAOL* (system architecture optimization language) [Kugele and Pucea 2014]. It is intended to easily but yet concisely specify both objectives and constraints, whereby special focus is placed on simplicity and usability by engineers rather than mathematicians.

Preconditions

Therefore, the engineer has to specify goals, constraints, and orders using SAOL. Orders are used to rank solutions according to the user's priority and to specify compatibility relationships (e.g., DAL classification). Moreover, an enriched model with deployable entities, be they software components (tasks, applications, etc.) or hardware components (e.g., field sensors, actuators, pumps, etc.) and a technical model of the deployment targets (e.g., ECUs, DIOs, inverters, etc.) has to be given. This means that at least the attributes of the modeling artifacts that are referred to within objectives and constraints have to be given.

Outputs

The method returns a number (depending on the selected algorithm) of valid deployments — that is, only those that satisfy all the given constraints.

Postconditions

The methods provided guarantee that only valid solutions are returned. The solutions can be persisted within the model for further processing.

Method description

Through its generic architecture, the *system architect* provides a flexible extension of the solver infrastructure. Currently, several solvers and combinations of them are supported:

❑ *MOEA (multi-objective evolutionary algorithm)*: the NSGA-II algorithm used returns "optimized" solution(s).
❑ *IDE (iterative deployment enumeration)*: an SMT-based approach iteratively enumerates (not optimized) deployment solutions.
❑ *MGIA (modified guided improvement algorithm)*: MGIA computes Pareto-optimal solution(s) for a multi-objective problem.
❑ *MOEA + IDE*: MOEA uses a valid IDE-generated initial population.
❑ *MOEA + MGIA*: MOEA uses an optimal MGIA-generated initial population.

Tool support

The Eclipse-based tool *System Architect* provides powerful modeling support: a text-based editor with syntax highlighting and intelligent text completion is accompanied by modeling support for all viewpoints of the SPES XT modeling framework.

9.4 Application to the Automotive Example

In the exterior lighting system and the speed control system described in Chapter 2, optimization is advantageous for several aspects. The available task structure has to be deployed to the available hardware resources. There are several variations for the hardware structure with some freedom in the selection and replacement of hardware components. Here, the method described in Section 9.3.1 can be used to select the appropriate hardware. For the deployment of tasks to the hardware architecture and the feasibility analysis, the second method explained in Section 9.3.2 may be used.

Ensuring certain timing constraints plays a crucial role for the optimization. Constraints on deadlines, activation periods, and execution demands force the distribution of the task structure to several hardware components. In the exterior lighting example, there are several timing requirements for the system which need to be validated.

All of the techniques described in Section 9.3.3 are suitable for considering timing properties, some of them using abstraction metrics on the utilization. Even if there is only a fixed hardware structure and deployment of software tasks to the hardware components, optimization may still prove beneficial. Most of the components have more than one task deployed to them. The scheduling of tasks on each resource can require optimization to achieve responsiveness and meet the (end-to-end) timing constraints of the system. Some techniques support static scheduling with others also supporting dynamic scheduling. For multi-core systems, this scheduling optimization might include a static assignment of tasks to different cores. One technique also allows separate optimization of scheduling parameters such as priorities.

9.5 Summary

This chapter focused on the SPES XT modeling frameworks methods to foster optimal deployment. During the design of software-intensive embedded systems optimizations of different natures occur. To support engineers during this design process, a methodological framework was presented. Different methodological building blocks along with their implemented techniques fit well into this generic framework, facilitating a seamless but yet flexible design of software-intensive embedded systems. To take optimization into account, several extensions to the SPES modeling framework were presented. The building blocks were classified and the application of the DSE method was illustrated using a case study from the automotive domain.

9.6 References

[Barret et al. 2009] C. Barrett, R. Sebastiani, S. Seshia, and C. Tinelli: Satisfiability Modulo Theories. In: Handbook of Satisfiability, IOS Press, 2009, pp. 825-885.

[Barret et al. 2010] C. Barrett, A. Stump, C. Tinelli: The SMT-LIB Standard: Version 2.0. In: Technical Report, Department of Computer Science, The University of Iowa, 2010.

[Büker et al. 2011] M. Büker, W. Damm, G. Ehmen, A. Metzner, E. Thaden, and I. Stierand: Automating the Design Flow for Distributed Embedded Automotive Applications: Keeping Your Time Promises, and Optimizing Costs, Too. In Proceedings of the International Symposium on Industrial Embedded Systems (SIES), 2011, pp. 156-165.

[Büker et al. 2013] M. Büker, W. Damm, G. Ehmen, S. Henkler, D. Janssen, I. Stierand, and E. Thaden: From Specification Models to Distributed Embedded Applications: A Holistic User-Guided Approach. SAE International Journal of Passenger Cars- Electronic and Electrical Systems, 2013, pp. 194-212.

[Chapman 2006] R. Chapman: Correctness by construction: a manifesto for high integrity software. In: Proceedings of the 10th Australian Workshop on Safety Critical Systems and Software, Vol. 55, 2006, pp. 43–46.

[de Moura and Bjørner 2008] L. de Moura and N. Bjørner: Z3: An Efficient SMT Solver. TACAS, LNCS. Springer, Berlin Heidelberg, Vol. 4963, No. 24, 2008, pp. 337–340.

[Diebold et al. 2014] P. Diebold, C. Lampasona, S. Zverlov, S. Voss: Practitioners' and Researchers' Expectations on Design Space Exploration for Multicore Systems in the Automotive and Avionics Domains: A Survey. In: Proceedings of the 18th International Conference on Evaluation and Assessment in Software Engineering, ACM, New York, 2014.

[Eisenring et al. 2000] M. Eisenring, L. Thiele, E. Zitzler: Conflicting Criteria in Embedded System Design. In: Design & Test of Computers, Vol. 17, No. 2, 2000, pp. 51-59.

[Hall and Chapman 2002] A. Hall and R. Chapman: Correctness by construction: developing a commercial secure system. Software IEEE, Vol. 19, No. 1, 2002, pp. 18–25.

[Hamann et al. 2006] A. Hamann, M. Jersak, K. Richter, R. Ernst: A Framework for Modular Analysis and Exploration of Heterogeneous Embedded Systems. Real-Time Systems, Vol. 33, No. 1-3, 2006, pp. 101-137.

[Hilbrich and Dieudonné 2013] R. Hilbrich, L. Dieudonné: Deploying Safety-Critical Applications on Complex Avionics Hardware Architectures. Journal of Software Engineering & Applications, Vol. 6, No. 5, 2013.

[Hofstedt and Wolf 2007] P. Hofstedt, A. Wolf: Einführung in die Constraint-Programmierung Grundlagen, Methoden, Sprachen, Anwendungen. Springer, Berlin Heidelberg, 2007.

[Kirkpatrick et al. 1983] S. Kirkpatrick, C. D. Gelatt Jr., M. P. Vecchi: Optimization by Simulated Annealing. Science, Vol. 220, No. 4598, 1983, pp. 671-680.

[Kondeva et al. 2013] A. Kondeva, D. Ratiu, B. Schatz, S. Voss: Seamless model-based development of embedded systems with AF3 Phoenix. ECBS, IEEE, 2013, pp. 212.

[Kugele and Pucea 2014] S. Kugele, G. Pucea: Model-Based Optimization of Automotive E/E Architectures. In: 6th International Workshop on Constraints in Software Testing, Verification, and Analysis, 2014, pp. 18-29.

[Rosinger et al. 2014] M. Rosinger, M. Büker, R. Weber: A User-Supported Approach to Determine the Importance of Optimization Criteria for Design Space Exploration. In: Proceedings of IDEAL'14 Workshop, IFIP Springer Series, Springer, Berlin Heidelberg, 2014.

[Rosinger et al. 2015] M. Rosinger, M. Büker, R. Weber: An Approach to Guide the System Engineer during the Design Space Exploration Process. In: W. Zimmermann, W. Böhm, C. Grelck, R. Heinrich, R. Jung, M. Konersmann, A. Schlaefer, E. Schmieders, S. Schupp, B. T. Widemann, T. Weyer (Eds.): Software Engineering Workshops 2015 (SE-WS 2015) - Gemeinsamer Tagungsband der Workshops der Tagung Software Engineering 2015, Vol. 1337, CEUR-WS.org, 2015, pp. 81-90.

[Thaden 2013] E. Thaden: Semi-Automatic Optimization of Hardware Architectures in Embedded Systems. PhD thesis, Carl von Ossietzky University of Oldenburg, 2013.

[Weber et al. 2014] R. Weber, S. Henkler, A. Rettberg: Multi-Objective Design Space Exploration for Cyber-Physical Systems Satisfying Hard Real-Time and Reliability Constraints. In: Proceedings of IDEAL'14 Workshop, IFIP Springer Series, Springer, Berlin Heidelberg, 2014.

[Wolf 2006] A. Wolf: Object-Oriented Constraint Programming in Java Using the Library firstCS. WLP, 2006, pp. 21–32.

Arnaud Boyer
Bastian Tenbergen
Santiago Velasco
Pablo Antonino
Peter Battram
Suryo Buono
Bernhard Kaiser
Justyna Zander
Kai Bizik
Alexander Prohaska
Michael Käßmeyer

10

Modular Safety Assurance

Most embedded systems in the automotive, avionics, or automation domains are safety-critical systems which are subject to strict safety standards and regulatory guidelines which govern the development process. These standards and guidelines require a tight integration between the development process and safety assurance. In modular engineering approaches such as the SPES XT modeling framework in particular, this means that safety assurance must be modular and enable developers to reuse partial safety results in different phases or between developments. However, without the support of sophisticated methods and tools, this approach is very tedious, error-prone, and costly. In this chapter, we introduce the SPES XT modeling framework's support for modular safety assurance. We first introduce the challenges that arise from the modularization of safety assurance and present the Open Safety Model, which provides the modularity and the compositional properties needed in order to achieve modular safety assurance within the SPES XT modeling framework. We also present methodological building blocks which support modular safety assurance during specific phases of development.

© Springer International Publishing AG 2016   169
K. Pohl et al. (eds.), *Advanced Model-Based Engineering of Embedded Systems*,
DOI 10.1007/978-3-319-48003-9_10

10.1 Introduction

In recent decades, at pace with the growth of safety-critical applications, the subject of functional safety has become more significant
within the development of embedded systems. Safety-critical systems must ensure that the occurrence of unwanted dysfunctional
events remains below an acceptable probability and does not lead to
harmful consequences for humans or external systems. Therefore,
safety assurance is concerned with identifying possible hazards of a
system and conceiving safety measures. These safety measures must
be suitable for avoiding or controlling systematic failures and for
detecting and mitigating random hardware failures.

As implied by safety standards [ISO 26262, ARP 4761], the safety process applies to the development of a system and ends with the
release of the system for wide usage. However, in common practice,
existing systems are often modified (either during operation or in
the form of re-releases), updated with new functionality or changes
in design, and system components are frequently reused in other
development processes (see Chapter 11). In these cases, safety
standards make it necessary to repeat the entire safety assurance
process for each component and the system as a whole in order to
be certain that the modifications do not impair the system's safety.
As a consequence, everything that has already been done in terms of
safety assurance has to be revised, without any allowance for the
safety assurance of parts that have been left untouched in the system, which in turn leads to a significant increase in development
time and cost.

It makes sense, therefore, and is desirable, to extend modular
development approaches, such as the original SPES modeling
framework, in order to allow modular safety assurance. Regardless
of how the specific development process using the SPES XT modeling framework is executed, an approach is needed which uses the
artifacts produced in the different viewpoints and in the different
granularity layers to assure the system's safety. Hazard and safety
analyses must take the artifacts of the SPES XT modeling framework into account such that as integration takes place during development, safety assurance results can be aggregated in-step with
development. We present such an approach in the remainder of this
chapter.

10.1.1 Overview of Modular Safety Assurance Activities

Fig. 10-1 gives an overview of the steps involved in a modular safe- *Modular safety process*
ty process for a system and its subsystems — as required by safety
standards — using the example of the safety-related work products
of the [V-Model 2015]. The inner track of the V-Model shows the
standard development activities which produce the artifacts of the
SPES XT modeling framework. The outer track shows the functional safety activities which are the focus of this chapter.

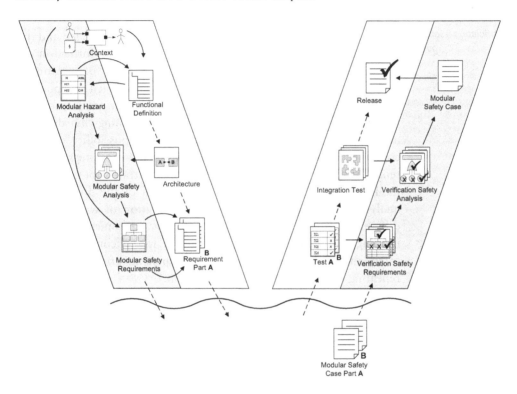

Fig. 10-1   *Process-oriented overview of the approach*

As the initial safety process step — the hazard analysis — requires a   *Hazard analysis requires*
multitude of information about the system, all safety activities start   *a multitude of*
with the definition of the scope of the system under development   *information*
(SUD). In this step, information is gathered about the purpose, preliminary architecture, and operational context [Daun et al. 2012].
In addition, other knowledge sources may provide supplementary
information that has an impact on the system safety [Daun et al.
2014]. Based on the scope of the system and the system's principle

functionalities, modular hazard analyses are conducted. The purpose of hazard analyses is to identify operational conditions of the SUD's functionality that could lead to harm. In modular development, hazard analysis results must be portable, meaning that it must be possible to reuse them in another context. Doing so requires explicit modeling of all assumptions about the usage context (see [Daun et al. 2014], [Daun et al. 2015]). The main outputs of such an analysis are hazards and safety goals (i.e. top-level safety requirements which are incorporated into the system's requirements artifacts [Tenbergen et al. 2015]).

Risk of violating a safety goal

Once the top-level safety requirements have been defined and the system design has been decided, safety analysis is performed in order to examine which system failures bear the risk of violating a safety goal or, in other words, can be a hazard. When components of the system originate from different suppliers or are intended to be reused, these analyses must be modular, meaning that different acceptable analysis techniques must offer the possibility of being combined into one comprehensive analysis for the integrated system. Therefore, the proposed approach concentrates not only on the modularization but also on the combination of the diverse safety analysis techniques, which enables a separation of work between different companies. In particular, whenever a technical component is being modified, the need to modify or repeat component-based safety analysis can be limited to the components affected instead of repeating the analysis of the entire system.

Define countermeasures

The next step in the safety process is to define countermeasures. In the automotive domain, the artifact explaining this derivation process is referred to as the *functional and technical safety concept* [ISO 26262]. The result of this step is a set of safety requirements which are superimposed onto the existing system requirements. The output resulting from this modularized activity is a set of independent safety requirement modules that can be allocated to individual components. A suitable way of achieving this modularity in safety requirement decomposition and assignment is the use of contracts.

Safety case

In the end, the fulfillment of all safety-related duties has to be demonstrated in a final report usually referred to as a safety case, showing the lines of arguments and the particular evidence for each claim therein. Accordingly, an efficient modular safety process will have to provide the means for composing this safety case step by step from independent parts originating from different contributors, perhaps partly reused from former projects. When a part of the system is affected by some later change, those parts of the safety case which have to be revalidated must be clearly identifiable; the remaining parts should be left untouched. Since the fulfillment of each safety requirement must be proven, a modular safety case benefits from modular safety analyses and modular derivation and assignment of safety requirements.

10.1.2 Challenges for Modular Safety Assurance

The safety analysis activities are integrated in the standard development processes of each industry domain and must be tailored to a specific context described, for example, in a safety plan. The proposed approach focuses on the following core challenges: *Integration of safety analysis activities*

❑ How can a hazard analysis in the early phase of the system development be modularized so that it can be reused whenever system components are reused?

❑ How can the safety analysis techniques be modularized? And how can the safety analysis techniques be heterogeneously combined with one another?

In addition, a safe system must also exhibit safety properties and safety requirements which ultimately need to be verified either by analysis or by test. These safety properties and safety requirements are often broken down and disseminated among the whole development without considering modularity or reusability. As a consequence, the following challenges arise: *Verification of safety properties and requirements*

❑ How can the safety requirement specification and the process of systematic construction of these requirements be modularized?

❑ How can the validation and verification of the safety artifacts be supported?

10.2 Integrated Safety Framework

One of the most important goals addressed by the modular safety assurance approach is expressed by the need to facilitate the reuse and certification process of safety-critical systems. The strategy chosen to achieve this goal is to follow a modular realization of the safety-relevant artifacts. The proposed approach relies on an ontological integration of different building blocks, which allows consolidation of a domain-independent safety aspect in order to support the development of safety-critical systems. This framework is also referred to as the *Open Safety Model* (OSM). Before we look at *Open Safety Model*

OSM, however, let us recall Section 3.4, which introduced the integration of crosscutting extensions within the SPES XT modeling framework. As depicted in Section 3.4, the safety aspect is presented as a perspective of the SPES XT modeling framework, in compliance with the definition of architectural perspectives in [Rozanski and Woods 2005].

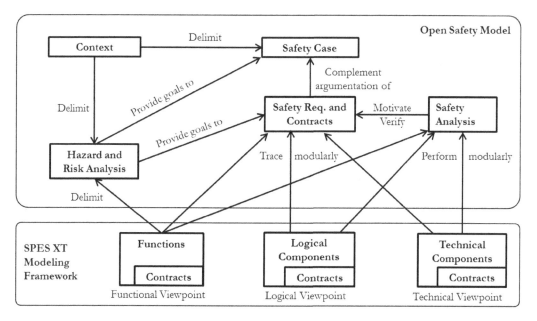

Fig. 10-2   *Open Safety Model entities and relationships*

OSM consists of modular building blocks

As we can see in Fig. 10-2, the OSM consists of a number of modular building blocks. These are described in Section 10.3. The following is a brief overview of OSM which illustrates the relationships between the modular building blocks in order to support the overall approach.

❑ The modeling of the safety-relevant *context* of a system is vitally important to assess and assure a system's safety for two essential reasons: on the one hand, a system receives input from other systems and human users and this input must not lead to hazardous situations [Tenbergen et al. 2014]. For example, if an external system provides faulty input or fails to provide input altogether, the SUD must maintain a safe state. On the other hand, whether or not a state is considered safe largely depends on the available information, such as hazard or safety analyses conducted by the stakeholders, or the engineering artifacts that have been developed so far. We use the SPES XT context modeling framework from Chapter 4 to document development and solution constraints (through context of knowledge models, see [Daun et al. 2014]) and to document assumptions about the functional interaction (through functional operational context models, see [Daun et al. 2015]), as illustrated in Section 10.3.1.

❑ As a mandatory element of safety engineering, a *hazard and risk analysis* is an assessment of the system to be developed [ISO 26262, ARP 4761] which aims to identify and classify hazards. The functions modeled in the functional viewpoint form a basis for hazard analyses because potential hazards are identified for each function or malfunction respectively [Tenbergen et al. 2014]. The results of hazard and risk analyses can necessitate new safety requirements or refine existing safety requirements which specify mitigations for each identified hazard or a reduction of its associated risk. Similarly, to safety requirements, hazard and risk analyses influence safety goals which are included in the safety case. Section 10 will discuss modular hazard analyses.

❑ *Safety analysis* is an umbrella term comprising various established failure analysis techniques such as *component fault trees* (CFT), *failure mode and effect analysis* (FMEA), and *Markov chains* [Ericsson 2005]. To allow the techniques to be executed in a modular way enabling reusability, the different analyses correspond to specific functions and components which are modeled in the functional, logical, and technical viewpoints. Similar to the hazard and risk analysis, the results of safety analyses are captured as safety requirements, as outlined in Section 10.

❑ *Safety requirements and contracts* document the mitigation of hazards and failures. The safety requirements are fundamental artifacts for the safety assessment of a system. To this end, safety requirements must be specified consistently and as formally as possible. Therefore, the approach uses various formalizations of safety contracts which explicitly state the assumption about the system's context and a guaranteed behavior of the system if the assumption is met. Contracts are hence one way to specify safety requirements which have been identified during hazard and risk analyses or safety analyses and are specified for functions, logical components, or technical components. More detailed information can be found in Section 10.3.3.

❏ A *safety case* constitutes an argument about the safety of a system [Kelly 1998]. This argumentation is done in a structured way focusing on the dependencies of the OSM entities involved. Therefore, a safety case comprises information from the safety context about the circumstances under which the safety argumentation holds, the hazards that have been identified and what risk they entail, and what mitigations have been proposed and documented to ensure safe operation. Section 10.3.3 illustrates the application of safety cases in the *Goal Structuring Notation* [GSN 2011].

Definition of a series of checks

With the help of these methodological building blocks, we can define a series of checks to validate the completeness and consistency of models. For example:

❏ All identified failure causes have traceable safety requirements.
❏ All identified functions have been considered during the hazard and risk assessment.
❏ All identified contracts potentially written with different formalisms are consistent with each other.
❏ All identified evidence in the safety cases reference an engineering artifact in the context of knowledge.

10.3 Methodological Building Blocks

This section describes the building blocks introduced in the previous section in more detail.

10.3.1 Safety Context Modeling

Safety aspects in context models

When discussing the safety of a system at the property level, no statement can be provided in absolute terms as the safety strongly depends on the context the system is deployed in (see [Daun et al. 2015], [Tenbergen et al. 2015]): the functional hazards that can occur during operation of the SUD and which depend on the interactions with other systems as well as human users — that is, the operational context. This is particularly important when discussing a system's safety in a modular way: because the safety argument of the entire system depends on the safety properties of its individual components, a description of the operational context in which a component is known to be sufficiently safe is vital in order to properly define system-level hazards and adequate safety measures.

The context modeling framework from Chapter 4 can be used for this purpose, as summarized in Tab. 10-1.

Tab. 10-1   *Safety context modeling building block*

SPES XT process building block	
ID	10.1
Name	Safety context modeling
Method	Modeling safety-relevant context information for reference in safety-related artifacts, for example: safety concepts, safety cases, or hazard analyses
Input	Optional: existing context models Optional: existing engineering artifacts of external systems
Output	Context models extended with safety-relevant information
Condition /Limitation	-

The SPES XT context modeling framework from Chapter 4 proposes four types of context models. In the following, we illustrate how context modeling can be used to support safety assessment activities.

Safety *context of knowledge* models provide a knowledge landscape for a modular safety case of the SUD, also called the *context subject*. As such, they supply the safety-relevant grounds which underlie the rationale for discussing the system's safety within the boundaries of its operational context. The main benefit of documenting this information is to ensure repeatability and ascertain the ability of the assumption upon which the safety argument is built. To do so, context of knowledge models document, for example, which safety-relevant engineering artifacts have been used as input for hazard analyses, who conducted the hazard analyses, what process guideline this person followed, where hazard analyses results were documented, and also which physical properties are relevant for the hazard analyseis (e.g., the road friction coefficient). The context of knowledge hence serves as a knowledge base which guides the modular safety assurance process, equivalent to a safety plan. Furthermore, context of knowledge models can be used to scope the safety case and structure evidence used in sub-safety cases, for example, by documenting how the mitigation for a hazard is partitioned into subsystems of the SUD. Hence, the context of knowledge models can also guide the modular composition of a safety case. The context of knowledge models have to be maintained over the entire duration of the system's development. In the event of a context change, the impact on the safety assurance has to be evaluated. If the system needs to work in a different operational

Safety context of knowledge models

context, the reusability of the safety context of knowledge could be easily checked.

Example of a context of knowledge model

Fig. 10-3 shows an example of a context of knowledge model for the adaptive cruise control system from Chapter 2. Safety-related information is provided for a set of hazards — for example, for the assurance that distance and velocity measurement is sufficiently accurate in terms of the physical properties of vehicles to be implemented. For further details, tolerances are provided for the distance and velocity respectively. As evidence for the absence of the documented hazard, a formal proof is planned for every implementation. The system is realized in compliance with ISO 26262 and addresses functional safety. As exemplified, the safety context modeling enriches the documentation for the context subject by giving comprehensive information about hazards to be mitigated during the realization process.

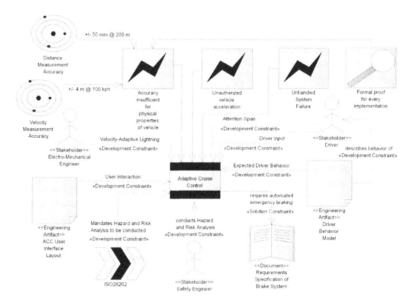

Fig. 10-3   *Safety context of knowledge of the adaptive cruise control (ACC)*

Structural operational safety context models

Structural operational context models (see Section 4.2.2) document which interactions take place between the SUD and external systems or human users. In other words, these models provide a static-structural representation of the external interfaces between the SUD and its operational context. These models can be used by the safety engineer as an additional source of information to guide the hazard analysis. For example, the safety engineer can use these models to

ascertain whether or not the hazard analysis is complete by systematically considering non-functional failures at each interface (e.g.,
thermal, electrical, mechanical..). In addition, these models can
provide clues about which external systems may be affected by malfunctions of the SUD and the operational situation involving external systems or human users may produce trigger conditions for a
hazard. This is the case when erroneous input from one system
causes the SUD to fail. Structural operational context models can
hence provide a completeness criterion for hazard analysis.

Functional operational context models focus on conceptual functions and services offered by external systems which the SUD can
consume, or functions offered by the SUD that are accessed by human users or external systems. In a similar manner to the structural
operational context models, the functional operational context
models are used by the safety engineer as an additional source of
information to complement the hazard analysis. These models
adopt a functional view of the interaction between the SUD and its
context — for example, in order to study the failure propagation
through a network of connected dependent systems. Furthermore,
functional operational safety models can be used to guide analysis
of unwanted side-effects as well as hazardous emergent behavior
(see Chapter 8), as shown in Fig. 10-4. To pick up the given safety
example for Fig. 10-4, the sensor measurement tolerances addressed
can be associated to functionality — for example, *Measure Relative
Velocity* — implemented for the ACC. A functional operational
context model can therefore discuss possible inconsistencies regarding context elements associated with the context subject in further
detail and at an early stage. In addition, this functional view can
stimulate the discovery of undetected hazards in the context of
knowledge model — for example, the influence of weather conditions on the sensors used.

*Functional operational
safety context models*

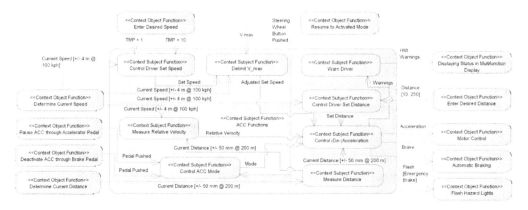

Fig. 10-4   *Functional operational safety context of the ACC*

*Behavioral operational
safety context models*

*Behavioral operational context model*s focus on externally observable states of entities and functions identified in the structural operational context and the functional operational context respectively. This is relevant for safety mostly in combination with the hazard analysis because both functions and entities within the context can be analyzed with regard to their externally observable states, thereby investigating how these externally observable states might cause hazards for other entities or functions. In hazard analyses, these externally observable states of external systems, human users, and functions represent conditions that, possibly in conjunction with other conditions, may lead to a hazard. The behavioral operational context can therefore be used to document the operational situation for hazard analyses. For example, the identification of a specific state which triggers a hazard may have an impact on decreasing the exposure of the hazard to the system being in that specific state. An example of the behavioral operational context depicting the states of a driver that an adaptive cruise control can observe is shown in Fig. 10-5.

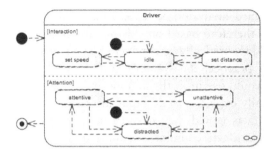

Fig. 10-5   *Behavioral operational safety context of the ACC*

10.3.2 Modular Safety Analysis

The first set of core challenges identified in the introduction is addressed in this section.

Modifications to Support Context Modeling

The identification and assessment of hazards constitute a mandatory activity in the field of safety engineering. This activity is conducted early in the safety process as its outcome is the foundation for all further activities in the safety life cycle. In the automotive industry domain, the objective of the hazard analysis and risk assessment (HRA) is to identify and categorize the hazards triggered by malfunctions of the SUD and to formulate the safety goals related to the prevention or mitigation of the hazardous events in order to avoid unacceptable risk. In the aerospace domain, functional hazard analyses (FHA, see [ARP 4761]) pursue identical objectives, with the aim of identifying relevant failure conditions and expressing the associated top-level safety requirements.

Composable hazard analysis

Tab. 10-2   *Modular hazard analysis modeling building block*

SPES XT process building block	
ID	10.2
Name	Hazard analysis modeling
Method	Modeling the hazards of a system in context-oriented, portable manner
Input	Functional models (e.g., item description), context models
Output	Safety-relevant hazards, top-level safety requirements, safety-relevant context
Condition/ Limitation	-

Modular hazard analysis (see Tab. 10-2) is an analysis building block and is therefore based on existing artifacts. Even though the method could theoretically be conducted on any granularity layer of the SUD, in practice, the system's granularity layer with human interaction would use it: in the automotive domain, an item definition is typically expected as an input in order to start a HRA; the required artifacts would then have to be taken from the available definition of artifacts contained in the requirement and the functional viewpoint (see Chapter 3). In the aerospace domain, the functional definition of an avionics system would have to be found at the exact same place.

Both HRA and FHA comprise a number of items of information which must be documented across the domains and related to one another. This information includes:

- ❑ Functions or subfunctions
- ❑ Malfunctions or failure modes
- ❑ (Hazardous/failure) scenarios
- ❑ Effects (including person at risk, safe state..).
- ❑ Hazards (hazardous events/failure conditions)
- ❑ Derived requirements

General approach to modeling hazard analyses   Fig. 10-6 depicts the general approach of the method for modeling hazard analyses based on the example of the ACC from Chapter 2. It consists of two generic activities: hazard identification and risk assessment.

Fig. 10-6   *Modular hazard analysis approach for the ACC*

Hazard identification   The hazard identification ensures that all hazards caused by malfunctions are collected in a systematic manner. This step is conducted on the basis of existing design artifacts which may either describe the system itself (see Chapter 3) or its operational context (see Section 10.3.1). For example, it can be appropriate to look for abnormal behavior of an interfaced system and undertake design decisions based on the consequences. Reciprocally, it is mandatory to know the impact of an abnormal behavior of the system in its operational context. In each case, a hazard is collected and modeled as a hazardous scenario artifact. It is important at this point (for the sake of the portability) to know which aspect from the context of knowledge influences the hazardous scenario: this information can be included in the model. Other conventional means for hazard identification — such as brainstorming, checklists, or quality history data — may also be appended to the safety context of knowledge for the purpose of the hazard identification.

Risk assessment focuses on the hazards identified in the previous step which are to be avoided in order for the system to be safe: in the aerospace domain, these hazards are failure conditions which could lead to danger for the crew, the passengers, or the airplane. In the automotive domain, these hazards could also pose a risk for persons in and around the vehicle (e.g., pedestrians or cyclists). During risk assessment, hazards are classified with domain-specific characteristics, such as:

Domain-specific risk assessment

❑ The *severity* of the hazard
❑ The *probability of exposure* to harm based on the individual operational situation of the system (e.g., the driving situation in the automotive domain)
❑ The *controllability* of harm and hazard, that is, the possible reaction of the system, human users, or external systems to prevent the hazard or prevent a resulting accident

In the automotive domain, these characteristics serve as the basis for the determination of the *automotive safety integrity level* (ASIL) for the system or item which may give rise to the hazard. According to [ISO 26262], there are four different ASILs, where ASIL A is the least strict safety integrity level and ASIL D the strictest. In the example shown in Fig. 10-6, the ASIL is defined as ASIL C.

In the aerospace domain, the FHA is applied similarly (but focusing on the severity of the hazards) to determine the *development assurance level* (DAL) that is needed in order to cope with systematic faults (see [ARP 4754A]).

As in the hazard identification, the risk assessment can refer to the context of knowledge, which may harbor relevant hazard scenarios. This information can also be included in this model.

Modifications to Support Composition

Composable safety analysis

Safety analysis plays a vital role in system safety assurance. For a given system safety context, we have to prove that the probability of all identified hazards is acceptably low. Therefore, a safety analysis of the system has to be conducted (see Tab. 10-3).

Tab. 10-3   *Modular safety analysis building block*

SPES XT process building block	
ID	10.3
Name	Modular safety analysis
Method	Modeling and analyzing a system's failure behavior using different safety analysis techniques in a component-oriented fashion
Input	Logical or technical architectural model
Output	Failure propagation model, safety analysis results
Condition/ limitation	Requires proper selection of the safety analysis techniques

There are two categories of safety analysis techniques: modular and non-modular techniques.

Modular techniques

Modular techniques can be used at any position inside the failure propagation model of the system. One example is a *component fault tree* (CFT) [Kaiser et al. 2003], which is a modular component-based version of the widely used fault trees (FT) and fully compatible with generic fault tree analysis [Veseley 1981]. Another example is an *interface-focused failure mode and effect analysis* (IF-FMEA) [Papadopoulos et al. 2001], which defines external causes and local effects for incoming and outgoing failure modes. In general, a modular integration into the safety analysis framework is possible if the technique supports the underlying principle of failure logic modeling (FLM), meaning that it can process incoming failure modes on the one hand and outgoing failure modes on the other hand.

Limits of modularization

However, not all safety analysis techniques can be fully modularized. Some, for example, can only describe systems which are independent of external failure modes. Hence, artifacts of such a technique cannot be integrated if failures can be triggered by the failure of other modules. A prominent example of this is Markov analysis [IEC 61165]. The integration of incoming failure modes for state-based analysis can be considered by using other more expressive modeling techniques such as *state event fault trees* [Kaiser 2005] or *Petri nets* [Peterson 1977]. However, to keep the analysis in the framework simple, the artifacts of Markov analysis are integrated as a source of failure modes to other modules in the system.

The failure model can be seamlessly integrated by connecting the output failure interfaces of one module with the input failure interfaces of other modules. This can easily be done due to the fact that the failure model abstracts from the realization of the respective safety analysis. This supports distributed and independent development of system components. The failure behavior of the whole system is typically composed bottom-up by establishing how failures propagate from one module to another. This information is necessary to trace failure modes from the failed module all the way up to the system boundary or to a mitigating module in between.

Seamless integration of failure model

The failure propagation is demonstrated by taking the adaptive cruise control (ACC) of the automotive system cluster (see Chapter 2) as an example. For the purpose of demonstration, the system is reduced to one radar sensor, the radar unit, and the ESP controller (see Fig. 10-7). Bold arrows indicate the failure propagation between components. The radar sensor has been modeled using a Markov chain in the lower left corner of Fig. 10-7. Failure mode *Sensor unavailable* is propagated to the radar unit and used as failure input inside the CFT. Due to the structure of the CFT, the failure mode directly triggers the failure *Braking command omission* of the radar unit. Since this failure mode is used as an external failure mode in the IF-FMEA of the ESP controller, the failure mode *Braking omission* is triggered. The impact of the radar sensor failure can be derived directly from the failure propagation model (right side of Fig. 10-7).

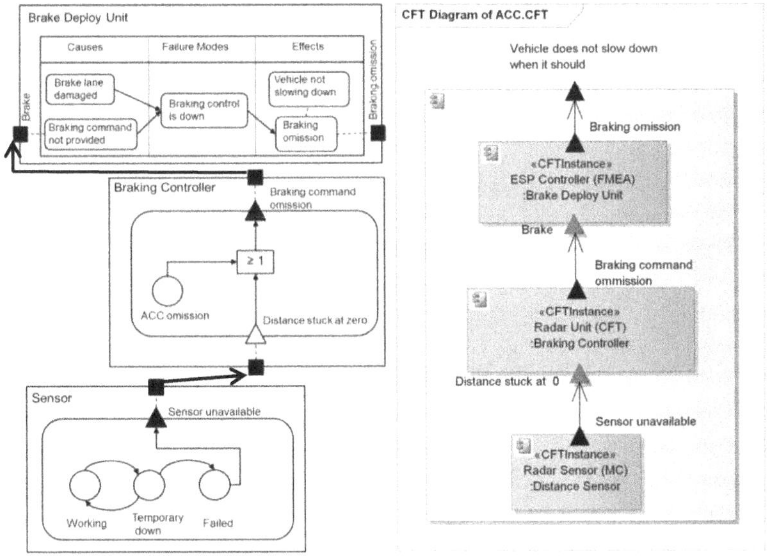

Demonstration of failure propagation

Fig. 10-7   *Modular failure propagation in the ACC example*

Qualitative and quantitative analysis

One main goal in performing safety analysis is to identify failures or combinations of failures leading to hazardous situations. The framework presented uses *minimal cut sets* (MCS) [Veseley 1981] to identify such combinations. This technique was developed for FTA and can be applied to CFTs without any restrictions. To realize a continuous qualitative analysis of heterogeneous models, FMEA model parts are transformed into CFTs by converting internal causes (IC) to basic events, failure modes (FM) to OR gates, external causes (EC) to input failure modes, and local effects (LE) to output failure modes (See Fig. 10-8a). Measures (M) are used for modeling means to avoid or mitigate any of these FMEA model parts, are transformed into AND gates. Markov models can also be integrated in a similar way; a set of states can be combined using an OR gate to form the output failure mode (See Fig. 10-8b).

Calculation of top-level hazard

The probability of the top-level hazard (e.g., total system failure) can be calculated directly from the list of MCS. This may lead to a conservative approximation of the failure probability. As a requirement for the quantitative analysis, all IF-FMEA models have to include information to quantify the probability of each internal cause identified. Note that the transformation from the MCS calculation is only a valid approximation if the transition probabilities between the states are very low. Otherwise, Markov models have to be transformed into fault tree instances containing only a single output failure mode and basic event before the top-level probability can be calculated. The results from qualitative and quantitative safety analysis are used to create new safety requirements or refine existing safety requirements respectively.

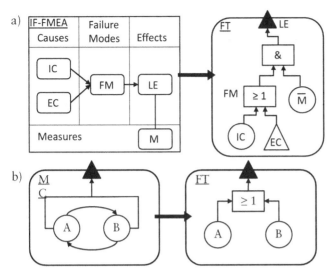

Fig. 10-8   *Examples of transformations from (a) IF-FMEA and (b) Markov to FT*

10.3.3 Safety Requirements and the Modular Safety Case

In this section, we address the second set of core challenges from the introduction.

Decomposition of Safety Requirements

Safety requirements are fundamental artifacts in the specification of safety-critical systems because they document how the hazards and failures of the system are mitigated. They are produced by hazard and safety analyses and may therefore come from various perspectives and levels of abstraction. In order to keep the safety requirements consistent, integrated, and modular, the notion of a safety requirements module is proposed. The role of these modules is to aggregate safety requirements. A description of the building block is given in Tab. 10-4.

Safety requirements module

Tab. 10-4   *Safety requirements module artifact building block*

SPES XT process building block	
ID	10.4
Name	Safety requirements module
Method	Safety requirements modules bundle the safety requirements of a SUD
Input	Top-level safety requirements from the hazard or safety analyses
Output	Decomposed safety requirements in the form of safety requirement modules that enhance modularity and reuse
Condition/ limitation	The complexity of the modules

Safety requirements module

A module defines the relevant safety requirements for a SUD at a particular granularity layer. Since safety requirements are not constrained to individual viewpoints or models therein, a safety requirements module can pertain to multiple models of multiple viewpoints. Within a module, the safety requirements are further decomposed according to the *Safety Requirement Decomposition Pattern* (SRDP). An example is given in the next figure.

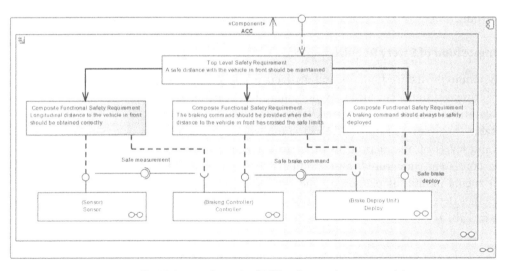

Fig.10-9   *Example of ACC's safety requirements module*

Composition of Requirements Using Contracts

Contract-based modeling

The safety requirements can be specified using various formalisms. The use of contract-based modeling may be beneficial in this regard.

Contracts represent a requirement in a structured way, separating it into an assumption which states the expected properties of the environment, and a guarantee which describes the desired behavior of the component provided the assumption is met by the environment. This separation is the foundation for building a sound theory that allows formal consideration of the composition of systems. The description of the building block is given in Tab. 10-5.

Tab. 10-5   *Contract-based safety requirement artifact building block*

SPES XT process building block	
ID	10.5
Name	Contract-based safety requirements
Method	Formalization of safety requirements by means of safety contracts to foster modularity and reuse
Input	Safety requirements
Output	Modular safety requirements (safety contracts)
Condition/limitation	-

There are three different levels of formality in the proposed approach:

❑ *Informal* contracts have the advantage that they can be read like natural language and are thus easy to understand, even for non-experts. However, they are often ambiguous and difficult to check. Informal safety contracts are used to specify the interaction between safety requirements modules.

❑ A *semi-formal* contract representation provides a predefined syntax that enables an engineer to formulate requirements in an unambiguous and more structured way that is still intuitive. Semi-informal safety contracts are used to specify the safety-relevant interface contracts in the architecture of a system.

❑ *Formal* contracts also have an underlying formal representation like automata. This formalism enables complex analysis such as consistency or dominance checks. Formal safety contracts are used to formally specify the safety requirements module of a component.

Semi-formal interface safety contracts

The relationship between assumptions of a specific component to guarantees of its neighbors (and vice versa) can be defined by an *interface contract*, where assumptions refer to signal qualities provided at input ports of a component and guarantees refer to signal qualities provided at the output ports of the neighbor components.

An interface contract is a combination of an assumption and a promise that are attached to the interface. Each contract is negotiated between two respective neighbor components. Assumptions and guarantees may refer to signal characteristics at a given port, as seen by an omniscient external observer — for example, the accuracy of a signal with respect to the real physical quantity, the integrity of a signal, or the delay of a port signal with respect to an event in the outside world. It is obvious that some reusable component, taken out of its environment, will not refer to such context knowledge that is not directly related to quantities observable at its direct interfaces — therefore, the full extent of modularization is only achieved by later transformation of interface contracts into component contracts. However, interface contracts help to verify the architecture when decomposing the system because the architecture can easily be checked along the signal flow links regardless of whether or not the assumptions at the input of some component are fulfilled by the guarantees at the linked output of its predecessor component. Unfulfilled contracts — that is, assumptions that are not satisfied by corresponding guarantees — can be detected and highlighted automatically, which has been demonstrated by a prototype tool in [Sonski 2013]. Fig. 10-10 illustrates the concept of the interface contract.

Formal component safety contracts
Component contracts are contracts between any SUD and its operational context. Therefore, they are a keystone in making (safety) requirements allocated to a component modular and reusable, and in making formerly tacit assumptions explicit. In the SPES XT modeling framework, the aim of component safety contracts is to consider the fault containment properties of components in a modular, reusable way [Oertel et al. 2014] and to conduct safety contract validation on the system's architecture. Such validation can serve as evidence for the correct allocation of safety functions which in turn mitigate hazards (cf. [Tenbergen et al. 2015]).

Interface contracts and component contracts can be transformed into each other canonically.

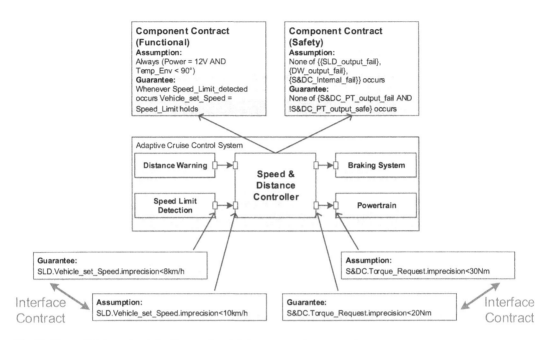

Fig. 10-10   *Safety contracts for the speed and distance control component*

Fig. 10-10 depicts an excerpt of the architecture of the adaptive cruise control system and two exemplary contracts linked to the *Speed & Distance Controller* component. The functional contract states, for example, that assuming that the component's power supply is sufficient and the temperature of the component's environment is suitable, the component will set the set speed to the speed limit when it is detected. A safety contract focusing on failure propagation states, for example, that the output of the *Speed & Distance Controller* component will not fail and will not be unsafe if the component's inputs or the component itself does not fail. These contracts are represented in a formal way using the predefined RSL (requirement specification language) patterns introduced in Chapter 6.

Architecture excerpt of the adaptive cruise control system

Composition of Evidence Using GSN

The goal of a safety case is to use a structured argument to demonstrate that the SUD is free from unacceptable risk. The modular safety case is special in the sense that it is less concerned about the safety of individual architectural elements but more about the interrelationships between the elements of the Open Safety Model (see

Safety case modeling

Section 10.2). It therefore takes the results of other safety-related methods as input (see Tab. 10-6) and shows in a clear and traceable manner how these results contribute to the overall safety of the system.

Tab. 10-6   *Safety case modeling building block*

SPES XT process building block	
ID	10.6
Name	Safety case modeling
Method	Modular safety case creation
Input	Evidence from all other safety-related methods; context of knowledge
Output	Structured argumentation about the system's safety using the *Goal Structuring Notation* (GSN)
Condition/limitation	-

Current industrial practice

In current industrial practice, a safety case is mostly understood as a collection of safety-related documents. While each document may have reached a sufficient level of maturity, it is hard to make justified assertions about the consistency and completeness of the documents' contents. The modular safety case uses a graphical notation enriched by trace links to elements of the Open Safety Model. This helps to determine and limit the effort needed for recertification: if an element in the OSM changes, we can directly determine which parts of the safety case are affected. The modular safety case also enables the derivation of consistency and completeness arguments.

Argumentation elements

The *Goal Structuring Notation* (GSN) [Kelly 1998] has proven useful and versatile for graphically documenting a system's safety argument in the past. The GSN is a graphical modeling language for creating structured arguments. An argument noted in the GSN consists of two types of elements, of which the inferential elements constitute the core of the argument:

❑ Goals are claims which form the argument. In many cases, they are motivated by the results of modular hazard analyses or safety analyses and may overlap with safety goals in the requirements viewpoint. Goals may be supported by (refined into) subgoals.

❑ Strategy elements may be used to describe the nature of the inference between a goal and its supporting sub-goals. This often expresses the rationale of a refinement and may be the implementation of an argumentation pattern as described later.

❑ Each goal has to be supported by evidence for the case to be convincing. Solution elements support goals and provide links to evidence items.

The second type of model element is the contextual element. While not a mandatory part of the argument, these elements provide valuable additional information.

❑ Contextual elements link to additional knowledge sources such as stakeholders, engineering artifacts, or regulatory documents in the safety context of knowledge that influence the argument [Daun et al. 2014].
❑ Justifications may be used to give a statement of rationale — for example, to justify exceptions from expected procedure.
❑ Assumptions are intentionally unsubstantiated statements. They specify under which circumstances the safety argument is valid. As assumptions significantly weaken the argument, they should be transformed into goals and be substantiated with evidence whenever possible.

The GSN community standard [GSN 2011] provides more information on the elements and the construction of a basic argument.

Standards such as [ISO 26262] prescribe evidence from verification and validation activities such as review, test, and simulation reports. These activities are described in a safety plan and the resulting work products are added to the safety context of knowledge. The results of the methods described in this chapter — for example, the verification of formal contracts — are also included in this way. In addition, the safety case provides arguments about the various methods' consistency and completeness. *Sources of evidence*

The GSN uses natural language, which is appropriate for creating convincing and understandable arguments. However, justified assertions about the consistency and completeness of the case require a certain level of formality. The modular safety case therefore extends the basic GSN by adding internal and external trace links. Fig. 10-11 shows an example argumentation fragment for the adaptive cruise control system from Chapter 2. *Internal and external trace links*

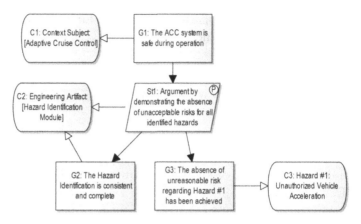

Fig. 10-11   *Example argumentation fragment using the GSN and trace links*

Some contextual elements in this example contain external trace links (denoted by square brackets) to items in the safety context of knowledge. In the example, the context elements C1 and C2 contain a link to the context subject *Adaptive Cruise Control* and the engineering artifact *Hazard Identification Module* respectively (see Fig. 10-6). This enables automatic completeness and consistency checks, for example, if all evidence elements connected to a context subject have also been considered in its corresponding safety case.

The example also demonstrates the use of internal trace links. A typical strategy is to iterate over all identified hazards. For each hazard, a sub-argument is developed to show that the hazard has been sufficiently mitigated or to justify why no further measures are necessary. The elements forming such a strategy pattern are connected by internal trace links. The existence of these links is denoted by the decoration (P symbol) in the upper right corner of the strategy element in Fig. 10-11. The pattern can then be checked for completeness and consistency. The results of these checks can in turn be used as indicators and evidence for the maturity of the safety case itself.

10.4 Summary

In this chapter, we described the SPES XT modeling framework's approach for supporting the safety engineers in overcoming the challenge of modular safety assurance. The proposed solution provides concrete answers to the questions raised in Section 10.1. The approach relies on a set of modular building blocks, such as composable hazard and safety analysis techniques, a structured and modular derivation of safety requirements, and an incremental safety case construction, which allows cost-efficient system modification and component reuse with regard to the accompanying safety obligations. The building blocks have been described extensively in Section 10.3.

In order to gain the most benefit from them, the building blocks are integrated into a combined framework called the *Open Safety Model*, which was briefly described in Section 10.2.

As an engineering challenge, modular safety assurance should continue to grow in the future as its related use case — the reusability of safety artifacts from existing systems — becomes more frequent in the industry. The mutual consideration of variability management (Chapter 11) together with the modular safety assurance approach could constitute a basis for promising future research theses.

10.5 References

[ARP 4754A] SAE International: ARP 4754A – Guidelines for Development of Civil Aircraft and Systems, 2010.

[ARP 4761] SAE International: ARP 4761 – Guidelines and Methods for Conducting the Safety Assessment Process on Civil Airborne Systems and Equipment, 1996.

[Daun et al. 2012] M. Daun, B. Tenbergen, T. Weyer: Requirements Viewpoint. In: K. Pohl, H. Hönninger, R. Achatz, M. Broy (Eds.): Model-Based Engineering of Embedded Systems: The SPES 2020 Methodology. Springer, Berlin Heidelberg, 2012.

[Daun et al. 2014] M. Daun, J. Brings, B. Tenbergen, T. Weyer: On the Model-Based Documentation of Knowledge Sources in the Engineering of Embedded Systems. In: Proceedings of the Fourth Workshop on the Future of the Development of Software-Intensive Embedded System Development (ENVISION2020), 2014, pp. 67-76.

[Daun et al. 2015] M. Daun, B. Tenbergen, J. Brings, T. Weyer: Documenting Assumptions about the Operational Context of Long-Living Collaborative Embedded Systems. In: Proceedings of the 2nd Collaborative Workshop on Evolution and Maintenance of Long-Living Software Systems (EMLS), 2015, pp. 115-117.

[Ericsson 2005] C. A. Ericsson II: Hazard Analysis Techniques for System Safety. Wiley 2005.

[GSN 2011] GSN Community: GSN Community Standard Version 1. Origin Consulting Limited, York, 2011.

[IEC 61165] IEC: IEC 61165:2006 – Application of Markov techniques. Edition 2.0, 2006.

[ISO 26262] ISO: ISO 26262-1:2011 – Road vehicles -- Functional safety -- Part 1: Vocabulary. 2011.

[Kaiser 2005] B. Kaiser: State/Event Fault Trees: A Safety and Reliability Analysis Technique for Software-Controlled Systems. PhD thesis, Technische Universitaet Kaiserslautern, Fachbericht Informatik, 2005.

[Kaiser et al. 2003] B. Kaiser, P. Liggesmeyer, O. Mäckel: A new component concept for fault trees. In: Proceedings of the 8th Australian workshop on safety-critical systems and software, Canberra, Australia, 2003, pp. 37-46.

[Kelly 1998] T. Kelly: Arguing Safety – A Systematic Approach to Managing Safety Cases. PhD thesis, Department of Computer Science, The University of York, 1998.

[Oertel et al. 2014] M. Oertel, A. Mahdi, E. Böde, A. Rettberg: Contract-based safety: Specification and application guidelines. In: Proceedings of the 1st International Workshop on Emerging Ideas and Trends in Engineering of Cyber-Physical Systems (EITEC), 2014.

[Papadopoulos et al. 2001] Y. Papadopoulos, J. A. McDermid, R. Sasse, and G. Heiner: Analysis and synthesis of the behavior of complex programmable electronic systems in conditions of failure. International Journal of Reliability Engineering and System Safety, Vol. 71, No. 3, 2001, pp. 229–247.

[Peterson 1977] J. L. Peterson: Petri Nets. ACM Computing Surveys, Vol. 9, 1977, pp. 223-252.

[Rozanski and Woods 2005] N. Rozanski, E. Woods: Software Systems Architecture: Working with Stakeholders Using Viewpoints and Perspectives. Addison-Wesley, Upper Saddle River, NJ, 2005.

[Sonski 2013] S. Sonski: Contract-based modeling of component properties for safety-critical systems. Master Thesis. Hochschule Darmstadt, 2013.

[Tenbergen et al. 2014] B. Tenbergen, A. C. Sturm, T. Weyer: A Hazard Taxonomy for Embedded and Cyber-Physical Systems. In: Proceedings of the 1st International Workshop on Emerging Ideas and Trends in Engineering of Cyber-Physical Systems (EITEC), 2014.

[Tenbergen et al. 2015] B. Tenbergen, T. Weyer, K. Pohl: Supporting the Validation of Adequacy in Requirements-Based Hazard Mitigations. In: Proceedings of the 21st International Working Conference on Requirements Engineering: Foundations for Software Quality (REFSQ), 2015, pp. 17-32.

[Veseley 1981] W. E. Veseley: Fault Tree Handbook. US Nuclear Regulatory Commission, Washington DC, 1981.

[V-Model 2015] V-Model – Wikipedia, Committee, https://en.wikipedia.org/wiki/V-Model. (Accessed May 29, 2015).

Ina Schaefer
André Heuer
Michael Himsolt
Tobias Jäger
Tobias Kaufmann
Christian Manz
Reinhold May
Christian Reuter
Bernhard Rumpe
Holger Schlingloff
Sebastian Schröck
Christoph Schulze
Michael Schulze
Stephan Weißleder

11

Variant Management and Reuse

Variability management and reuse are important concerns in the development of variant-rich software-intensive systems. In this chapter, we present the SPES XT modeling framework's mechanism to capture the orthogonal concern of variability. We then devise building blocks for realizing variant management and planned reuse, as well as for the assessment of variable artifacts.

© Springer International Publishing AG 2016   197
K. Pohl et al. (eds.), *Advanced Model-Based Engineering of Embedded Systems*,
DOI 10.1007/978-3-319-48003-9_11

11.1 Introduction

Overview   Modern embedded systems are highly variable in order to meet varying customer requirements and environment contexts. This variability is a crosscutting concern during software and systems development and must be considered in every development phase and its corresponding artifacts. Variability must also be consistently captured over different development phases and artifact types in order to allow consistent derivation of system variants. Variability modeling is orthogonal to the modeling of the actual development artifacts. The original SPES modeling framework [Broy et al. 2012] does not consider variability. Because of the high variability of modern embedded systems, there is a need to extend the SPES modeling framework to support variability for specifying systems as well as their operating contexts and to express variability as a first class engineering concept.

However, the characteristics of different engineering domains with respect to variant management and reuse vary — for example, in the lot sizes or the organizational structure. While in some domains development is based on systematic reuse, there are other domains, such as automation and avionics, where this degree of systematic reuse is not common yet [Große-Rhode et al. 2013]. Hence, specific methodological building blocks have to be provided to allow reuse-based engineering and seamless variant management within existing tool landscapes and development processes.

Problem space variability (cf. [Czarnecki and Eisenecker 2000]) defines the variability of a system from an abstract perspective in terms of configuration options that can be selected to derive variants. Solution space variability (cf. Kang et al. 1990) captures the variation of the artifacts used to build the actual product variants. Problem space variability and solution space variability are connected by configuration knowledge which determines how a particular configuration of the problem space variability is realized by binding solution space variability.

In this chapter, we present an extension of the SPES modeling framework with an orthogonal variability modeling perspective. This variability modeling perspective is based on a variability modeling ontology that can be instantiated with different problem space variability modeling languages — such as feature modeling and orthogonal variability modeling — and includes different views on variability to support the principle separation of concerns. Furthermore, we provide methodological building blocks for reuse-based engineering which allow the identification or development of reusable components with variability and configuration of these components for specific product variants. In addition, we propose seamless variability management based on a variability exchange language supporting integration of existing development tools. To assess the quality of our variability-aware development processes, we provide metrics for quality in reuse-based development.

The structure of this chapter is as follows: we describe how the SPES XT modeling framework extends the original SPES modeling framework to support variant management and reuse. To do so, we introduce a distinct variability perspective in Section 11.2. In Section 11.3, we consider methodological building blocks for reuse-based development and variability management as well as for quality assessment of reusable artifacts. Section 11.4 concludes this chapter.

11.2   Variability Extension to the SPES Modeling Framework

In this section, we describe how to capture variability of the system under development (SUD) and how this variability can be structured by the use of different views. The view concept is applied to the variability of the context in which the SUD is to be deployed.

11.2.1 Variability Perspective

All SPES viewpoints are affected by variability because variability is a crosscutting concern during software and systems engineering. Hence, the SPES XT modeling framework provides a *variability perspective* (cf. Heuer et al. 2013). This variability perspective documents all variability information orthogonally to the SPES viewpoints. The concept of the variability perspective is depicted in Fig. 11-1. Variability information that originates from different SPES viewpoints and different granularity layers can be interconnected within the perspective. Variability information can also be related to base artifacts.

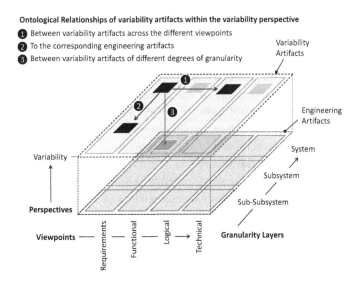

Fig. 11-1   *Variability perspective*

Variability perspective   The variability perspective defines the ontological concepts for variability modeling in accordance with [Pohl et al. 2005] and does not prescribe the use of certain variability modeling languages and is hence open to domain-specific instantiations. In order to capture variability as a separate concern, separate variability models (e.g., [Kang et al. 1990], [Clements and Northrop 2002], [Bachmann et al. 2003]) are based on two ontological concepts: the *variability subject* and the *variability object* and relationships between these two concepts. An overview of the variability modeling ontology used as a basis for the extension of the SPES modeling framework can be found in Fig. 11-2. Each ontological element is described in Tab. 11-1.

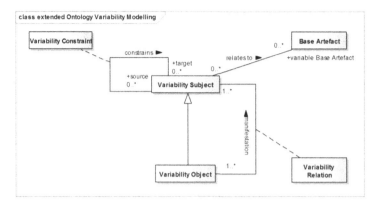

Fig. 11-2   *Variability ontology*

Tab. 11-1   *Description of variability ontology elements*

Element	Description
Variability subject	A *variability subject* is a variable item of the real world or a variable property of such an item and describes *what* can vary.
Variability object	A *variability object* is a particular instance of a variability subject which describes a specific variation of the variability subject. Hence, a variability object is always related to at least one variability subject via a variability relation.
Variability relation	A *variability relation* documents the relationship between variability subjects and variability objects.
Variability constraint	A *variability constraint* is a restriction between variability subjects. Due to the inheritance relationship between the variability subject and variability object, variability constraints can be defined between variability subjects and variability objects as well as between variability objects.
Base artifact	A *base artifact* is an engineering artifact that contains variability. Engineering artifacts are defined based on the SPES viewpoints.

The problem space variability modeling language to be used in the variability modeling extension of the SPES XT modeling framework must instantiate the ontology concepts defined in the variability perspective. In the following, two variability modeling languages are described which instantiate the ontological concepts of the variability perspective.

Problem space variability

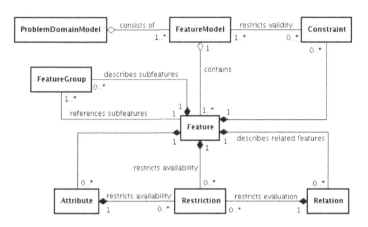

Fig. 11-3   *Feature model metamodel*

The metamodel of the variability modeling language for feature models (cf. Kang et al. 1990) depicted in Fig. 11-3 instantiates the ontological model of the variability perspective (cf. Fig. 11-2). The metaclasses *Constraint*, *Relation,* and *Restriction* are instances of the ontological metaclass *Variability Constraint*. Furthermore, the metaclass *Feature* is an instance of the ontological metaclasses *Variability Subject* and *Variability Object*.

Orthogonal variability modeling, feature modeling

Another possible instantiation of the variability ontology (cf. Fig. 11-2) is provided by the *Orthogonal Variability Model* (OVM) (cf. [Pohl et al. 2005]). In the OVM metamodel (cf. Fig. 11-4), the metaclass *Variation Point* instantiates the ontological concept *Variability Subject*. In contrast, *Variant* is an instantiation of *Variability Object*. The OVM metaclass *Variability Dependency* instantiates the ontological metaclass *Variability Relation*. The OVM metaclasses *Optional, Alternative Choice,* and *Mandatory* are also instances of *Variability Relation*. The ontological metaclass *Variability Constraint* has three different instances in the OVM metamodel. The first instance is *Variation Point Constraint Dependency* and describes constraints that are defined between variation points. The second instance is *Variation Point to Variant Constraint Dependency* and describes constraints between variants and variation points. The third instance is *Variant Constraint Dependency* and describes constraints between different variants.

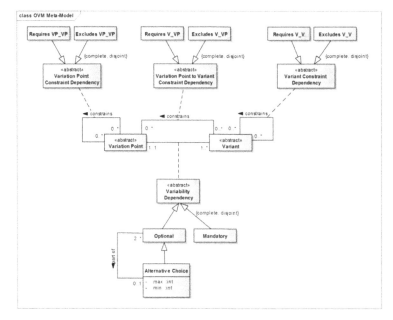

Fig. 11-4   *Meta-model for orthogonal variability modeling*

In the solution space, three main approaches for realizing variability are known (cf. [Schaefer et al. 2012]). First, annotative approaches represent all system variability in a single artifact. These variability models are also called 150% base artifacts because multiple variants can be derived by removing variability. Second, compositional variability modeling approaches realize different variants of a base artifact by extending it by variant-specific features.

Solution space variability

The third main approach is transformational variability modeling (cf. [Schaefer et al. 2012]), which describes variability in the form of incremental changes. In this direction, delta modeling assumes that a system family is described by a core model and a comprehensive set of deltas. The core system is usually a valid variant of the system family that only captures mandatory features. A delta [Clarke et al. 2011, Manhart et al. 2013] captures a set of cohesive changes to an existing core model and can be stored as an individual artifact. A specific variant is defined by the core model and a sequence of applicable deltas. Of course, constraints, such as the order of application for deltas and a pairwise exclusion, apply. Deltas have a practical advantage over annotative and compositional variability modeling: in industrial practice, we can start with an existing base model and successively extract deltas for existing projects, thus successively filling the library of available deltas.

Delta modeling

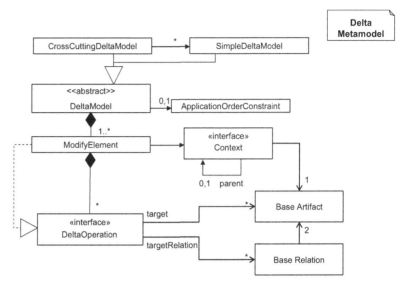

Fig. 11-5   *Metamodel for delta modelling*

Model extension to   To express variability in the solution space, the SPES XT modeling
express variability   framework allows for delta modeling (cf. Fig. 11-5). A CrossCuttingDeltaModel defines a set of SimpleDeltaModels which describe
modifications for different artifacts which belong to different viewpoints. Each DeltaModel defines a set of DeltaOperations which
transforms elements of the SPES XT modeling framework.

Based on this extension of the SPES XT modeling framework, we
can express variability with transformations captured in deltas. In
the example, the logical architecture of the advanced driver assistance system (ADAS) (cf. Section 2.2) is modeled by a core model in
Fig. 11-6. The DeltaModel in Fig. 11-7 modifies the ADAS module
by adding a *Limiter* module and corresponding ports (marked with
a +) and applying further modifications to the *Cruise Control* module. In addition, other delta models are necessary to modify models
of other viewpoints in order to obtain a consistent variant across
several viewpoints.

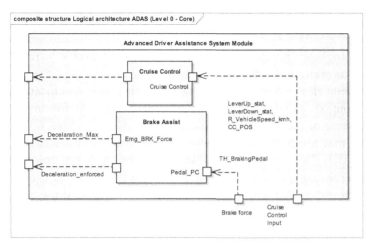

Fig. 11-6   *Core architecture model for ADAS*

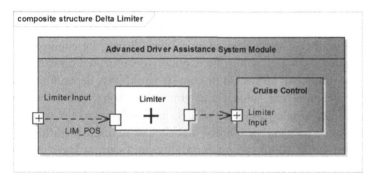

Fig. 11-7   *Delta limiter for ADAS*

11.2.2 Structuring the Variability Perspective

The variability perspective comprises the full set of variability in-   *Variability views*
formation that is relevant to the variable SUD. However, not all
variability information is relevant at a specific stage of the development of a system. This means that we have to structure the variability information. In order to do so, we introduce the concept of
variability views within the variability perspective. Variability views
follow the notion of IEEE Std. 42010:2011 (cf. [ISO/IEC/IEEE
42010]). A *variability view* is a work product which expresses the
variability information and related base artifacts with regards to
specific variability concerns. A specific variability view can be understood as a projection applied to variability information, the variability to the base artifact relationship, and the set of base artifacts.
All *variability concerns* that are addressed should be made explicit.

For instance, the variability of a SUD and the variability of the context of a SUD are variability concerns that can be used to structure the variability perspective. The concerns of a specific variability view as well as the concepts and rules for constructing the corresponding view can again be specified by a viewpoint specification.

Example: Context variability

Context variability refers to context entities — for instance, hardware of the SUD — which can be variable. This kind of variability may affect the variability of the embedded software as part of the SUD. Thus, context variability must be explicitly considered during the engineering of embedded software where context entities are given "as they are". It is not possible to change those entities (see Chapter 4), whereas the system itself can be formed. Context variability, therefore, cannot be changed, adapted, or added during the engineering of the SUD. Furthermore, context variability depends on the granularity layer considered. Hence, system variability on the granularity layer "System" may be considered as context variability on the subsequent lower layer.

Context variability handling

In order to handle context variability during the engineering of embedded software, it must be explicitly documented on each layer. As variability can be system or context variability, depending on the current engineering subject and granularity layer, it can be documented in a single variability model that conforms to the variability ontology. Views on variability models can be used to differentiate between context and system variability on a certain layer of a specific engineering subject under consideration.

Context and system variability interrelations

Context and system variability may have interrelations: specific variants in the context may require specific variants in the system and vice versa. These interrelations can be modeled, for example, by *requires* dependencies between model elements in the context variability view and the system variability view.

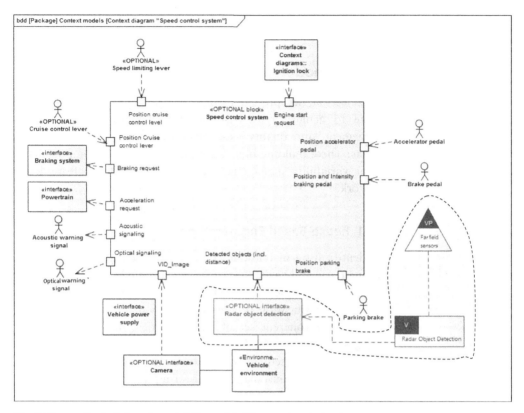

Fig. 11-8   Context view of a variability model of the speed control system

Fig. 11-8 shows the context view of a variability model and the logical architecture of the example in Chapter 2 of the speed control system on the system layer. The speed control system exists in different variants: it can include a simple cruise control or adaptive cruise control. An adaptive cruise control system needs radar object detection to detect vehicles ahead. However, this radar object detection is not part of the engineering subject (which is the software of the speed control system). Thus, the radar object detection is in its context. In addition, the radar object detection is not part of every vehicle that is delivered to a customer. Hence, it is variable and thus context variability is necessary for the current SUD on the current granularity layer. The radar object detection is required by the software of the adaptive cruise control. Accordingly, there is a *requires* dependency between the system variability (adaptive cruise control) and the context variability (radar sensor). More details can be found in [Kaufmann et al. 2014], [Heuer and Pohl 2014], and [Heuer et al. 2015].

11.3 Methodological Building Blocks

In this section, we present methodological building blocks which operate to achieve reuse-based engineering, seamless variability management, and quality assessment for the variable engineering artifacts. These building blocks consider the reuse of components as well as variability management based on the SPES XT modeling framwork.

11.3.1 Reuse-Based Engineering

Reuse-based engineering process

The identification of reusable components is the first step towards enabling reuse and variant management. Reusable components have to incorporate problem and solution space variability as previously described. Identification of reusable components is the process of determining a coherent set of functionalities that can be used in varying system contexts. The application of a reusable component constitutes the consequent integration into a different system. In order to establish reuse-based engineering in an organization, processes have to be adapted to integrate the process steps *Identification and development of reusable components* and *Application of reusable components* as illustrated in Fig. 11-9.

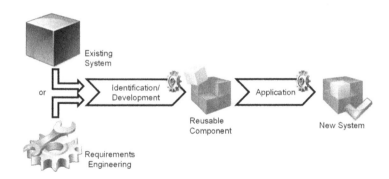

Fig. 11-9   *Reuse-based engineering process overview*

Identification and Development of Reusable Components

There are two ways of acquiring a reusable component: identifying a reusable component from within an existing system or developing a reusable component from a set of requirements for a required functionality. The process building blocks 11.1 and 11.2 reflect these options.

Tab. 11-2   Identification of configurable interdisciplinary reusable components

SPES XT process building block	
ID	11.1
Name	Identification of configurable interdisciplinary reusable components
Method	1. Analysis of existing life cycle artifacts of existing solutions with regard to the intended functionality of the reusable component 2. Extraction and adaptation of life cycle artifacts 3. * Modeling of variability according to the variability perspective 4. Preparation of the documentation, describing the reusable component and its possible application
Input	– Intended functionality of reusable component – Domain-specific artifacts from existing solutions
Output	– Reusable component (consisting of domain-specific artifacts and optional variability models), including documentation for reusable component
Condition/ limitation	The (partial) steps marked by (*) are only necessary if the reusable component shall include variability.

Within the engineering of automated plants or avionics systems, it is common to use existing solutions as a foundation for new systems. The instantiation of this process building block for the avionics domain focuses on safety, because safety is an important focus in avionics. In order to be safe, reusable components must serve a single purpose and artifact parts that do not directly contribute to that purpose must be removed. Safety is also ensured by extensive verification. Therefore, supporting verification artifacts within the reusable component is important.

Use of existing solutions

In order to develop reusable components, first, those parts of the life cycle artifacts that are only relevant to the original product in which the component was previously used must be removed. One of the outcomes of this step should be the reduction in the number of the requirements of the reusable component to its bare minimum. Next, a context analysis has to identify those life cycle artifact parts that depend on the context of the reusable component. On reintegration, the context-independent artifact parts of the component may be left unchanged, whereas the context-dependent artifact parts of the component have to be adapted to their new context. Finally, the reusable component is equipped with a description of

Development of reusable components

context-independent and context-dependent parts that eases integration into its new environment. An alternative instantiation of this idea for the engineering of automated plants can be found in [Schröck et al. 2015a].

Tab. 11-3   *Development of configurable interdisciplinary reusable components*

SPES XT process building block	
ID	11.2
Name	Development of configurable interdisciplinary reusable components
Method	1. Identification and analysis of the requirements for the reusable component 2. Development of a functional structure to fulfill the requirements 3. Development of a logical and technical solution 4. * Definition of variable parts following the variability perspective 5. Preparation of the documentation, describing the reusable component and its possible application
Input	– Requirements for reusable component
Output	– Reusable component (consisting of domain-specific artifacts and optional variability models), including documentation for reusable component
Condition/ limitation	The (partial) steps marked by (*) are only necessary if the reusable component shall include variability.

Systematic development of reusable components

In contrast to the identification of reusable components from existing systems in building block 11.1, building block 11.2 presents the systematic development of reusable components from scratch. This approach is chosen when the functionality that has to be developed is not yet available within the organization. The development of a reusable component from scratch is exemplified for the automation domain. The functions of a reusable component have to be defined by the discipline which dominates the early phases of the engineering process. In the case of a desalination plant (see Section 2.3), this is the process engineering discipline. The function definitions of this discipline have to be discussed with all other disciplines involved in the reuse approach. A reusable component within the automation domain contains artifacts described in Section 2.3, such as the piping and installation diagram (P&ID), circuit diagrams, and automation software designs (CFC/SFC), but is enhanced by variability models of the variability perspective. Further information is found in [Schröck et al. 2015a].

Application of Reusable Components

After the identification or development of reusable components, process building block 11.3 focuses on the engineering of the customer-specific solution using reusable components.

Tab. 11-4   *Engineering of a system using configurable interdisciplinary reusable components*

SPES XT process building block	
ID	11.3
Name	Engineering of a system using configurable interdisciplinary reusable components
Method	1. Identification of customer requirements for the system from a textual specification 2. Design of the functional structure of the SUD 3. Selection and configuration of the reusable components 4. Engineering of the parts that are not covered by reusable components; technical configuration of the reusable components 5. Consolidation of the solution 6. Documentation
Input	– Configurable interdisciplinary reusable components – Customers' specification
Output	– Artifacts of the SUD – Documentation for the SUD – Issues for potential development or adjustment of reusable components (optional)
Condition/ limitation	The type of system has to be well-known to the engineering organization (or must be assembled from well-known reusable components) otherwise no benefit could be expected from searching for reusable components.

Process building block 11.3 is useful if there are reusable components for parts of the system. In order to reduce risk, the method can be implemented incrementally by providing configurable reusable components for more important or frequently required components of the technical system first. There is no need to provide reusable components for all functions of the system in the first place [VDI 3695-2].

Incremental method implementation

Assuming that there are existing reusable components, the project-dependent method given in building block 11.3 can be applied to the engineering of a desalination plant example as it is described in Section 2.3. For example, reusable components include the beach well but also a seawater tank or the reverse osmosis membrane racks. The variability models used for the problem and solution space as well as the relationships between them allow a configuration of the reusable component that is mainly based on the requirements and only completed in detail by aspects of the implementation. By selecting and configuring the reusable components, the engineering of the desalination plant from the requirements analysis to the commissioning can be accelerated if suitable reusable components are provided. In addition, the deep knowledge embodied in the reusable components applied results in a more precise early cost calculation [Bramsiepe et al. 2012].

Prototypical method implementation

The aforementioned methods have been prototypically implemented within a tool chain for the engineering of automated plants as shown in [Schröck et al. 2015a] and [Schröck et al. 2015b]. The reusable components can be selected and configured by feature models that represent the problem space variability (see Section 11.2) within the tool pure::variants [Beuche 2013]. The configuration can be exported by pure::variants to be available for the engineering tool chain. As a prototypical interface between pure::variants and the Siemens tool chain, a simple CSV or XML file can be used. Within the solution space, an annotative approach for modeling the variability is used (see Section 11.2.1). These models were structured according to [ISO 10628-1], which also represents the guideline for the structure of the base objects within COMOS. Therefore, the import of the configuration only requires an allocation of the elements of the model representing the solution space and the base objects within COMOS. The import to a customer-specific project within COMOS generates the reusable component according to the configuration and parameterization that has been defined in pure::variants. These components have to be connected and, if necessary, further parts of the plant have to be developed. After the detailed engineering is finished, the data can be exported to subsequent software tools such as a process control system. Further information can be found in [Schröck et al. 2015a] or [Schröck et al. 2015b].

11.3.2 Variability Management

Variability management has to be incorporated into existing tool landscapes and development processes. To address this, in this section, we propose a generic variability exchange language and processes for seamless variant management. The generic variability exchange language allows the expression of problem and solution variability in a tool-independent manner.

Variability Exchange Language

Tools for variant management frequently use artifacts such as model-based specifications, program code, or requirements documents. This is often a two-way communication: variant management tools import variability information from an artifact and in return, export variant configurations. For example, tools need to gather information about the variation points that are contained in the artifact, need to know which variants are already defined, and then modify existing variants or define new variants or new variation points

There is currently no standardized API available for such operations. Hence, a variant management tool has to implement a separate interface — and possibly a new data format as well — for each new artifact. In the worst case, with m variant management tools and n artifacts, this may require the implementation of $m \times n$ different interfaces, as shown in Fig. 11-10.

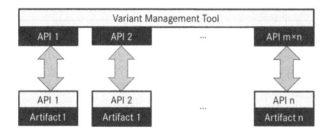

Fig. 11-10   *Without variability exchange language: many APIs have to be implemented*

The *variability exchange language* allows variant management tools to communicate with APIs of artifacts through a standardized interface, as shown in Fig. 11-11. If implemented across both variant management tools and artifacts, this reduces the number of required implementations significantly. A generic interface typically also lowers the barrier for adding new artifacts and supports the introduction of new tools for variant management.

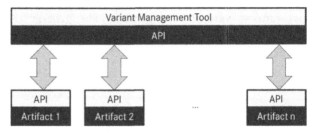

Fig. 11-11   *With variability exchange language: only a single API has to be implemented*

The *variability exchange language* provides a generic description of the variation points that are contained in an artifact. The variability exchange language is derived from the ontology presented in Section 11.2 and makes its classes and relations more specific towards real world applications.

Variation points may come in two flavors: *structural* and *parameterized*. *Structural variation points* are locations in an artifact which are removed or set inactive in a binding process. These variation points again fall into two categories: *optional* and *alternative*. This is implemented by defining a condition for each variation point.

Example 1: Disabling cruise control auto repeat

```
<optional-structural-variationpoint id="CC_no_autorepeat">
     <variation id="CC_Autorepeat_V">
            <condition type="or-feature-condition">
                   USA, Canada
            </condition>
     </variation>
</optional-structural-variationpoint>
```

Example 1 shows an optional variation point which controls auto repeat for cruise control (cf. Section 2.2). Some countries, for example the USA or Canada, do not allow auto repeat, so the variation point has a condition that checks for these countries.

Parameterized variation points

Parameterized variation points are values that are used to provide values for variables. Such variation points either choose one value out of several options or provide expressions which are evaluated in the binding process to assign a value to the parameter. Example 2 shows a parameterized variation point which chooses between an increment of 1 km/h and 10 km/h for cruise control (cf. Section 2.2).

Example 2: Variation points for cruise control increment

```
<xor-parameter-variationpoint id="CC_Increment">
    <variation id="CC_Increment_Fine">
            <condition type="single-feature-condition">
                    FineCruiseControlIncrement
            </condition>
            <value>1</value>
    </variation>
    <variation id="CC_Increment_Coarse">
            <condition type="single-feature-condition">
                    CoarseCruiseControlIncrement
            </condition>
            <value>10</value>
    </variation>
</xor-parameter-variationpoint>
```

Variation points may also define constraints and relations (see Section 11.2.1). For example, a set of variation points may be designated as alternatives, which means that all but one of them will be removed during the binding process. Furthermore, variation points may also contain other variation points which establish a hierarchical structure.

In addition to providing a generic description of variation points, the *variability exchange language* can define specific variant configurations. In our context, a variant configuration is an assignment of fixed values to the conditions or expressions that are associated with variation points. The binding process can further be assisted by assigning binding times to variation points, for example, `compile-time` or `postbuild-time`. The *variability exchange language* supports the declaration of multiple binding times per variation point; in this case, one binding time must be selected when the variation point is bound. This allows the decision about which binding time to use to be delayed.

Variant configurations

Finally, the *variability exchange language* defines standardized API operations for exporting and importing variability information. It is not expected that every API for an artifact or a variant management tool implements the full set of operations, hence the *variability exchange language* provides a way to state the capabilities of a specific implementation.

This leads to the following two methodological building blocks:

Tab. 11-5   *Import of artifact variability*

SPES XT process building block	
ID	11.4
Name	Import of artifact variability
Method	A variant management tool imports information on variation points from one or more artifacts. To make implementation easier and more versatile, we define a standardized exchange language and an API for variability information.
Input	Arbitrary SPES artifacts
Output	Document in *variability exchange language*
Condition/ limitation	-

Tab. 11-6   *Configuration of artifact variability*

SPES XT process building block	
ID	11.5
Name	Configuration of artifact variability
Method	A variant management tool defines product variants and exports them to the artifacts that realize the associated product line. To make implementation easier and more versatile, we define a standardized exchange language and an API for variability information.
Input	Document in *variability exchange language* Arbitrary SPES artifacts
Output	Arbitrary SPES artifacts (now configured for a specific variant)
Condition/limitation	-

Seamless Variant Management

Variant management of single artifacts such as requirements, functional models, or tests can be performed with existing methods and tools. Beyond this, seamless variant management addresses the consistent documentation, implementation, configuration, and traceability of variability over the complete application life cycle. Seamless variant management has to handle variability in diverse artifact types and representations (e.g., text-based and model-based) in parallel. To complicate issues, a given variation point can be implemented by different variability mechanisms resulting in different variant management solutions.

In the following, methodological building blocks for two fundamental aspects of seamless variant management are described. The building block templates rely on the variability perspective of the SPES XT modeling framework and have to be instantiated for the variant management process of each project.

Fundamental aspects of seamless variant management

To enable seamless variant management, the following prerequisites have to be satisfied: initially, variability has to be identified in the artifacts in the SPES XT modeling framework. The variability is subsequently extracted from this artifact and documented using the variability perspective (cf. Section 11.2.1) — for example, by importing it via the variability exchange language. This type of seamless variability model comprises a problem space variability model, such as a feature model, and at least one artifact-specific configuration model. A configuration model describes all variation points in an abstract form and relates to their implementation in a solution space variability model, that is, artifacts with variability. In addition, the configuration model defines configuration rules between selection of the features in the feature model and a variation point-specific value assignment.

The following methodological building blocks can be used if variability is added to a base artifact. The objective is to document all implicit variability knowledge in a seamless variability model.

Create/update seamless variability model

Tab. 11-7 *Create/update seamless variability model*

SPES XT process building block	
ID	11.6
Name	Create/update seamless variability model
Method	A stakeholder defines or updates a variation point in an artifact and documents the related abstraction of this variation point in a configuration model of the seamless variability model. Afterwards, configuration rules of the variation point have to be defined. As a result, any new product variability is documented in a seamless variability model independent of the SPES XT artifact and variability technique used.
Input	Variable SPES XT artifacts, (seamless variability model)
Output	(Updated) seamless variability model
Condition/limitation	-

Derive a specific artifact
variant
The following methodological building block can be used if a specific variant is derived. The objective is to use the seamless variability model and derive artifacts for a specific product variant.

Tab. 11-8   *Derive a specific artifact variant*

SPES XT process building block	
ID	11.7
Name	Derive a specific artifact variant
Method	A stakeholder defines a specific variant by selecting features of the seamless variability model. Using configuration rules, all related variation points are configured to realize the specific variant characteristics.
Input	Seamless variability model
Output	Variant-specific SPES XT artifacts
Condition/limitation	-

Illustration of
instantiation
A concrete instance of the methodological building blocks described can be illustrated by a simplified engineering method with the activity *requirements engineering*: in a change request scenario, a requirement engineer analyzes an upcoming change and adopts the corresponding requirement artifacts by adding a new alternative requirement. At this point, the whole set of artifacts is no longer consistent because the variability model does not reflect the new alternative requirement and the implementation realizes the corresponding alternative functionality. To make the system consistent again, the product line engineer executes building block 11.6 and documents the new variability in the variability model by adding a corresponding variation point. In addition, a new alternative in the functional model is implemented. After that process, the models are consistent again, and by executing building block 11.7, variant-specific requirements documents can be derived.

11.3.3 Assessment of Variability

In this section, we address the problem of how to evaluate software and system designs with respect to their potential for reuse and variability management. A classic dichotomy in the evaluation of software and systems is the distinction between process and product quality [Mishra and Schlingloff 2008]. In general, the quality of a product as the result of a design process is critically influenced by the quality of this process; however, even the best engineering processes cannot guarantee that the resulting products will have the required quality. Therefore, in practice, we have to continuously evaluate both the design processes and the resulting products. Since the SPES methodology focuses on product development, in the following we discuss which metrics are adequate for assessing variable artifacts. For practical use, these are to be integrated with the process assessment methods proposed, for example, in [Känsälä et al. 2005] and [van der Linden et al 2007].

In model-based development, typical model metrics can be used to measure model complexity: the number of model elements, their interconnections, the nesting depth of model elements, etc. However, these metrics are not very suitable for assessing the reuse potential. For this, metrics relating different model variants have to be applied. For example, the ratio of common and variant-specific elements can be used to assess the maturity of product portfolio scoping or to perform commonality and variability analyses. In the SPES XT modeling framework, we propose to evaluate the base model in relation to the product models in three aspects:

Measure complexity

1. *Similarity of alternative elements:* If there are several alternatives with a similar model structure, it might be possible to extract common elements for reuse. The ratio of common to different subelements produces a similarity metric.

2. *Number and scope of side effects:* Connections in a function model indicate dependencies caused by communication needs. A high number of dependencies among function component variants induced by function connections indicates that variance could potentially be better encapsulated. To obtain a concrete metric value, dependencies in a component model with variation points are counted using the connections whose endpoints are variability elements, that is, alternatives or options.

3. *Location of variation points in hierarchy levels:* Experience shows that a detailed commonality analysis leads to variation points that are located closer to the bottom of the component hierarchy in the model. The corresponding metric value is obtained by measuring the average distance of the variation points from the root of the component hierarchy.

Tab. 11-9   *Quality metric analysis for generic models*

SPES XT process building block	
ID	11.8
Name	Quality metric analysis for generic models
Method	An assessment method (with tool support) to evaluate the reuse potential of generic artifacts
Input	*Variability exchange language* document plus associated base artifacts
Output	A numeric assessment value and improvement possibilities within the input models.
Condition/limitation	-

Problem of evaluation of reuse potential

A problem with the evaluation of the reuse potential of models is that reuse is not universal but rather is bound to particular contexts or tasks. Reusability is not a property of a component on its own but of a certain environment in which the development is done. Thus, as products are generated in application engineering from artifacts designed in domain engineering, there must be a closed feedback loop from application engineering to domain engineering providing appropriate data on the actual reuse. Moreover, important product metrics such as the efforts for quality assurance of a reusable asset or the cost for certification using such artifacts can be obtained by measuring the associated processes.

11.4 Summary

Variability is prevalent in modern embedded systems. It is a cross-cutting concern that affects all phases of development and all associated development artifacts. In this chapter, we presented the SPES XT modeling framework's extension to deal with software and system variability within the application life cycle of variant-rich embedded systems. In addition to a variability perspective in the SPES XT modeling framework for capturing artifact variability, we devised different process building blocks in order to facilitate reuse-based engineering, seamless variability management, and assessment of reuse-based development products. The main challenges for variability management and reuse in the future are collaborative embedded systems which comprise a heterogeneous network of communicating nodes. As those systems constitute an open environment, there is a need to integrate the heterogeneous variability models in the problem space and the variability realization techniques in the solution space in order to guarantee correct operation of the resulting collaborative system variant.

11.5 References

[Bachman et al. 2003] F. Bachmann, M. Goedicke, Leite, J. Do Prado, R. Nord, K. Pohl, Balasubramaniam Ramesh, Alexander Vilbig: Managing Variability in Product Family Development. In: van der Linden, Frank (eds.): Proceedings of the 5th International Workshop on Product Family Engineering, (PFE-5), Volume 3014 of Lecture Notes in Computer Science, Springer, Berlin, Heidelberg, 2003, 66-80.

[Bramsiepe et al. 2012] C. Bramsiepe, S. Sievers, T. Seifert, G. D. Stefanidis, D. G. Vlachos, H. Schnitzer, B. Muster, C. Brunner, J. P. M. Sanders, M. E. Bruins, G. Schembecker: Low-cost small scale processing technologies for production applications in various environments – Mass-produced factories. Chemical Engineering and Processing: Process Intensification, Vol. 51, 2012.

[Broy et al. 2012] M. Broy, W. Damm, S. Henkler, K. Pohl, A. Vogelsang, T. Weyer: Introduction to the SPES Modeling Framework. In: K. Pohl, G. Böckle, F. van der Linden: Software Product Line Engineering: Foundations, Principles and Techniques. Springer, Berlin Heidelberg, 2005.

[Clarke et al. 2011] D. Clarke, M. Helvensteijn, Ina Schaefer. Abstract delta modeling. ACM Sigplan Notices, Vol. 46, No.2, 2011, pp. 13–22.

[Clements and Northrop 2002] P. Clements, L. Northrop: Software Product Lines – Practices and Patterns. Addison-Wesley, Boston, 2002.

[Czarnecki and Eisenecker 2000] K. Czarnecki, U. Eisenecker: Generative programming: methods, tools, and applications. Addison Wesley, Boston, 2000.

[Große-Rhode et al. 2013] M. Große-Rhode, P. Manhart, R. Mauersberger, S. Schröck, M. Schulze, T. Weyer: Anforderungen von Leitbranchen der deutschen Industrie an Variantenmanagement und Wiederverwendung und daraus resultierende Forschungsfragestellungen (published in German). In: Wagner, Lichter (Eds.): SE 2013 - Software Engineering 2013: Workshopband, Informatik (GI-Edition Proceedings, 215), 2013, pp. 251–260.

[Heuer and Pohl 2014] A. Heuer, K. Pohl: Structuring variability in the context of embedded systems during software engineering. In: Proceedings of the 8th International Workshop on Variability Modelling of Software-Intensive Systems (VaMoS '14), ACM, New York, 2014, pp. 21:1-21:8.

[Heuer et al. 2013] A. Heuer, T. Kaufmann, T. Weyer: Extending an IEEE 42010-Compliant Viewpoint-Based Engineering Framework for Embedded Systems to Support Variant Management. In: G. Schirner, M. Götz, A. Rettberg, M. C. Zanella, F. J. Rammig (Eds.): Proceedings 4th International Embedded Systems Symposium, Paderborn, 2013. Springer, IFIP Advances in Information and Communication Technology, 2013.

[Heuer et al. 2015] A. Heuer, T. Kaufmann, M. Constantinescu-Fomino: On the Explicit Consideration of Context Variability in the SPES Modeling Framework. In: W. Zimmermann, W. Böhm, C. Grelck, R. Heinrich, R. Jung, M. Konersmann, A. Schlaefer, E. Schmieders, S. Schupp, B. T. Widemann, T. Weyer (Eds.): Software Engineering Workshops 2015 (SE-WS 2015) - Gemeinsamer Tagungsband der Workshops der Tagung Software Engineering 2015, Vol. 1337, CEUR-WS.org, 2015, pp. 61-70.

[Kang et al. 1990] K. C. Kang, S. G. Cohen, J. A. Hess, W. E. Novak, A. S. Peterson: Feature-oriented domain analysis (FODA) feasibility study (No. CMU/SEI-90-TR-21). Carnegie-Mellon Univ Pittsburgh Pa Software Engineering Inst, 1990.

[Kaufmann et al. 2014] T. Kaufmann, C. Manz, T. Weyer: Extending the SPES Modeling Framework for Supporting Role-Specific Variant Management in the Engineering Process of Embedded Software. Software Engineering (Workshops), 2014, pp. 77-86.

[Manhart et al. 2013] P. Manhart, P. Nazari, B. Rumpe, I. Schaefer, A. Haber, C. Kolassa: First-Class Variability Modeling in Matlab/Simulink. In: Proceedings of the 7th International Workshop on Variability Modelling of Software-Intensive Systems, ACM, 2013, pp. 11–18.

[Mishra and Schlingloff 2008] S. Mishra, H. Schlingloff: CMMI Process Area Compliance with Formal Specification-Based Software Development. In: SERA 2008, Software Engineering Research, Management and Applications; Aug. 20-22, 2008, Prague, Czech Republic IEEE Computer Society Press, 2008.

[Pohl et al. 2005] K. Pohl, G. Böckle, F. van der Linden: Software Product Line Engineering: Foundations, Principles and Techniques. Springer, Berlin/New York, 2005.

Heidelberg, 2005. [Schaefer et al. 2012] I. Schaefer, R. Rabiser, D. Clarke, L. Bettini, D. Benavides, G. Botterweck, A. Pathak, S. Trujillo, K. Villela: Software Diversity: State of the Art and Perspectives. STTT Vol. 14, No. 5., 2012, pp. 477-495.

[Schröck et al. 2015a] S. Schröck, A. Fay, T. Jäger: Systematic interdisciplinary reuse within the engineering of automated plants. In: 9th Annual IEEE International Systems Conference (SysCon), IEEE, Vancouver, Canada, 2015.

[Schröck et al. 2015b] S. Schröck, F. Zimmer, A. Fay, T. Jäger: Systematic reuse of interdisciplinary components supported by engineering relations. In: 15th Symposium Information Control Problems in Manufacturing (INCOM), IFAC/IEEE/IFIP/IFORS, Ottawa, Canada, 2015.

[VDI 3695-2] VDI 3695-2, November 2010. VDI 3695 Part 2 – Engineering of industrial plants; Evaluation and optimization; Subject processes.

Part IV

Evaluation and Technology Transfer

Ulrich Löwen
Birthe Böhm
Alarico Campetelli
Maria Davidich
Florian Zimmer

12

Experiences of Application in the Automation Domain

In this chapter, we explain the application of the SPES XT modeling framework for the running example of the desalination plant. The objective is to explain the philosophy and methodology of the SPES XT modeling framework to automation domain experts based on artifacts, processes, and tools usually used today in the automation domain. For this purpose, we elaborate degrees of granularity and different viewpoints for the engineering of automation software of a desalination plant. The specific characteristics of the automation domain in the context of developing embedded systems thus become obvious for non-experts. We draw conclusions from the application of the SPES XT modeling framework for automation domain experts and give hints on how they can benefit from a systematic model-based approach.

© Springer International Publishing AG 2016
K. Pohl et al. (eds.), *Advanced Model-Based Engineering of Embedded Systems*,
DOI 10.1007/978-3-319-48003-9_12

12.1 Introduction

Software and me-
chanical engineering
and electrics are
mutually dependent

If we look more closely at the embedded software in the automation domain, such as for desalination plants, it is apparent that the software is heavily influenced by mechanical engineering and electrics and vice versa. For example, to optimize the technological process of desalination, mechanical units are added or modified. These changes in mechanical engineering influence the automation software controlling these mechanical units. This means that any changes in any part of the design of one of these three disciplines change the context of the design of the other disciplines.

Consideration of
interaction

As a consequence, all engineering approaches have to consider the interaction between mechanical engineering, electrical engineering, and automation software engineering. In addition to resolving dependencies between these disciplines, we also have to take into account that the environment of the automation domain is changing. The increasing complexity of industrial plants, the shortening time-to-market, and the need for improvements in efficiency in engineering mean that a purely sequential engineering of the different disciplines is no longer feasible. Therefore, a parallelization of engineering is necessary, which requires co-engineering of the different disciplines with well-defined synchronization points.

Problem statement

From a software engineering point of view, the artifacts used in the automation domain today typically do not clearly differentiate between the automation software as the system under development (SUD) (i.e., the automation software of the desalination plant) and the system under consideration (i.e., the entire desalination plant) as propagated by the SPES XT modeling framework. In addition, there is a mixing of terminology between the different disciplines involved. For example, usually there is no clear separation between a drive as a physical object, a drive as an electrical component, and the automation software for controlling the drive. In addition — conditioned primarily by the capabilities of the engineering tool used — there is no consequent differentiation between requirements, functions, and solutions. However, a thorough conceptual separation is an important precondition for applying the SPES XT modeling framework.

Approach

A common approach, which is becoming more and more established in practice in the automation domain, is to create interdisciplinary components which allow deep integration of the disciplines inside a module but also define a clear interface of the modules and therefore encapsulate the requirements, function, and solution inside the modules (see Chapter 10). It should be possible to reuse interdisciplinary components across different plant projects in order to tap into their full potential in terms of enhanced quality, time, and cost savings. However, such modules are often hard to derive because system boundaries and structures in mechanical engineering, electrical engineering, and automation software engineering are not aligned. As a consequence, interdisciplinary components are usually only successfully employed when the system boundaries of the different disciplines are identical [Fay 2009].

A promising approach to aligning the system boundaries of the different disciplines is to create interdisciplinary components based on a functional breakdown structure of the plant (see also Chapter 10). This functional breakdown structure must be independent of a technical solution but must consider all influencing aspects such as spatial structures, process technological dependencies, the scope of supply from a purchasing perspective, etc.

A model-based approach, like the SPES XT modeling framework has to address these challenges of the automation domain and provide suitable solution approaches.

12.2 Today's Process

Fig. 12-1 shows a typical engineering process for a desalination plant. The desalination plant itself is already described in Chapter 2.

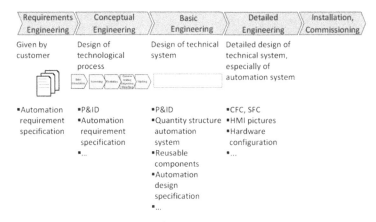

Fig. 12-1   *Engineering process for a desalination plant*

Engineering process

First, we consider the whole system in which the specific SUD (i.e., the automation software of the desalination plant) is embedded.

In general, the requirements for a desalination plant are handed over by the customer who places an order for the design, implementation, installation, and commissioning of a desalination plant. Hence, requirements engineering in the automation domain generally means capturing the customer requirements and clarifying open points. The conceptual engineering, which basically designs the technological process (e.g., the desalination process for the desalination plant), takes place based on these requirements. The most important resulting artifact in this process step is the piping and installation diagram (P&ID), which describes the technological process and the required mechanical equipment — see Fig. 12-2 for an example. Based on the design of the technological process, some disciplines such as mechanical engineering already start their design process and take decisions based on previous results. In basic engineering, subsequent disciplines such as automation software engineering and electrical engineering draft their later solutions. The more detailed design of the technical system — including engineering of the automation software — takes place during detailed engineering. Finally, the whole plant is installed and commissioned and the operation of the plant starts.

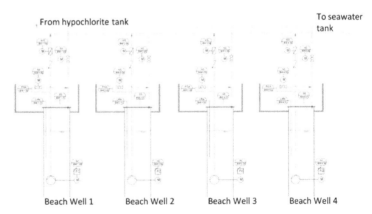

Fig. 12-2   *P&ID of the beach wells of a desalination plant (Siemens COMOS)*

12.3   Technological Hierarchy

The main input artifact for engineering the automation software is the P&ID diagram. In addition to the technological process and the required mechanical equipment, it implicitly includes a hierarchical structure which we call the technological hierarchy. Sometimes the technological hierarchy is more function-oriented with respect to the plant functions; sometimes it is more physically oriented with respect to the mechanical units. There are even some well-established standards, especially [ANSI/ISA-88.01-1995].

Fig. 12-3 shows an excerpt of a typical technological hierarchy of a desalination plant.

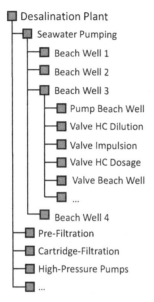

Fig. 12-3   *Excerpt from a typical technological hierarchy of a desalination plant*

The diagram in Fig. 12-2 illustrates the P&ID of the beach wells of the desalination plant which corresponds to the node *Seawater Pumping* in the technological hierarchy. Both Fig. 12-2 and Fig. 12-3 visualize that *Seawater Pumping* is composed of four beach wells: The technological hierarchy contains four nodes *Beach well 1..Beach well 4* (with equivalent sub-nodes not fully shown in this figure), while the P&ID visualizes four equivalent substructures, each of which show the design of one beach well.

With respect to the SPES XT modeling framework, the techno-   *Degrees of granularity*
logical hierarchy is used as a guide for the definition of the granularity layers. In general, the root node of the technological hierarchy is equivalent to the uppermost layer, while its child nodes belong to the next granularity layer and so on.

12.4 Applying the SPES Viewpoints in the Automation Domain

The SPES XT modeling framework described in Chapter 3 is based on four viewpoints which distinguish between the problem and solution space for a SUD. Note that in this chapter, the SUD is the automation software of the desalination plant, which covers the plant-specific software for controlling the behavior of the entire desalination. In the following, we illustrate the application of the SPES XT modeling framework to the seawater pumping section of a desalination plant.

12.4.1 Requirements Viewpoint

Requirements engineering starts with P&ID

Requirements engineering for the automation software starts with the P&ID and the technological hierarchy. Therefore, the requirements viewpoint can be structured according to the technological hierarchy, with the various requirements delivered by the customer or other disciplines then attached to the corresponding node in the requirements viewpoint. The extent to which the requirements should be detailed and assigned to the corresponding nodes of the technological hierarchy depends on the specific project and its constraints.

Requirements with respect to the entire desalination plant, which have an effect on the automation software of the desalination plant, are part of the context of the requirements viewpoint.

Identifying relevant components

Requirements engineering should then proceed with identifying the relevant components and interfaces between these components based on the available requirements. Fig. 12-4 shows components identified for the desalination plant automation software together with their input and output signals as they are mentioned in the requirements documentation and the P&ID, see also [Campetelli et al. 2015] for further details. In this example, these are the components *Seawater Pumping*, *GUI*, and *Seawater Tank*. The component *Seawater Pumping* comprises the automation of the beach wells which are themselves contained in the diagram in Fig. 12-4 as components *Beach Well 1...Beach Well 4*. The technological hierarchy shown in Fig. 12-3 also shows these relationships between *Seawater Pumping* and the beach well components. The SUD is represented by the components *Seawater Pumping*, *Seawater Tank*, and *GUI*, while the other components are the system's context. In the following, the SUD will be restricted to the component *Seawater Pumping*.

Thus, the system boundaries of the SUD are the input and output signals of the component *Seawater Pumping* shown as arrows in the figure.

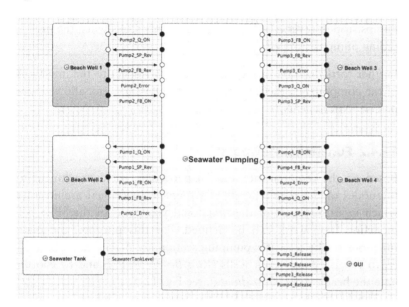

Fig. 12-4   *Identified components of the SUD "Seawater Pumping" and its context in the requirements viewpoint (AutoFOCUS 3)*

The next step is to detail the identified components (i.e., *Seawater Pumping*) of the SUD based on the given requirements. This means that the subnodes of the *Seawater Pumping* component are identified according to the technological hierarchy and the appropriate requirements are attached to the corresponding nodes in the requirements viewpoint.

The four beach wells collect the seawater at the shore and filter the water based on natural sand layers before the water is processed further. The P&ID describes the process and the required equipment and thus also describes requirements of the automation. Additional requirements of the automation include performance characteristics of the pumps and preconditions and postconditions of the interaction between the pumps and their environment as well as input and output signals of the beach wells.

Today, requirements engineering in the automation domain typically means the analysis and refinement of textual requirements, diagrams — mostly delivered by process, mechanical, and electrical engineering — or equipment lists. A typical text from a requirements specification by the customer could be:

Detail the identified components

> "Pumps of the beach wells start and stop cascaded.
> Pumps start in the following order: pump of beach well 1; pump of beach well 2; pump of beach well 3; pump of beach well 4. The next pump only starts when the preceding pump is running.
> Pumps stop in reverse order and a pump only stops when the preceding pump is stopped."

Further formalization of requirements with respect to the interfaces of the components, the introduction of a glossary, etc. allows more precise requirements.

12.4.2 Functional Viewpoint

Main functions are described

Based on the requirements and, in particular, their attachment to the nodes of the requirements viewpoint, a functional architecture which is already part of the functional viewpoint of the SPES XT modeling framework can be defined. For this purpose, the main functions of the seawater pumping section are described — basically, to run the four beach wells (*Run Beach Well_i*) and to coordinate the beach wells (*Coordinate*), see Fig. 12-5.

The functions are related to the requirements which are represented by a code (e.g., *BWx* for requirements related directly to the functionality of the beach wells and ASx for requirements concerning the beach well coordination by the automation system in Fig. 12-5). Both functions can be further detailed based on the requirements, see Fig. 12-7.

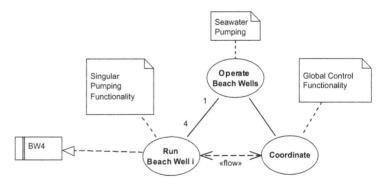

Fig. 12-5   *Functional architecture based on requirements categorization*

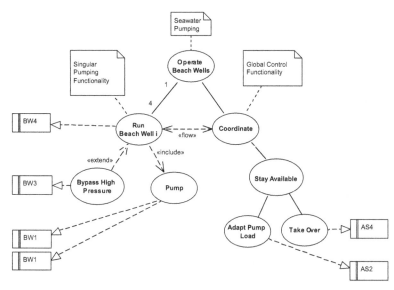

Fig. 12-6   *Detailed functional architecture shows a mapping of the requirements (AS and BW) to the functions*

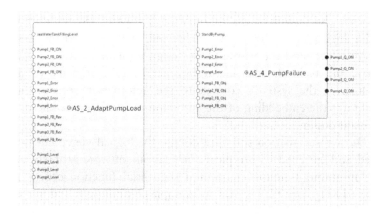

Fig. 12-7   *Functional architecture of the seawater pumping section*

A next step is to allocate these functions with the relevant interfaces of the components — that is, which inputs and outputs are used and generated by the functions. This results in a functional architecture as it is shown in Fig. 12-7. Functions are represented as boxes, inputs are integrated as circles on the left side of the boxes and outputs on the right side. The figure contains the function *Adapt pump load* which was already introduced in Fig. 12-6 but now its inputs and outputs are defined.

Allocate these functions

In addition to the functional architecture, state diagrams can be used to describe the dynamic behavior of the system. In turn, this can be used to validate the requirements (see also Chapter 6).

12.4.3 Logical Viewpoint

Derivation of logical architecture

The logical architecture of the automation software is derived based on the functional architecture. Typically, the structure of the logical architecture is guided by the technological hierarchy, meaning that the technological hierarchy is the leading design principle for the automation software design.

This type of leading design principle is reasonable when taking practical issues into account: the technological structure of the overall system in which the SUD — that is, the automation software — is embedded provides a stable structure for the system's context and the SUD. Even in the case of late changes during the engineering process, the technological hierarchy usually remains unchanged and thus provides a reliable basis for communication (e.g., between mechanical, electrical, and automation software engineering) and also for the design of the automation software. However, if the technological hierarchy has to be changed, then it is clear that it has severe effects and the whole plant design must be revised. In summary, the technological hierarchy ensures easy transfer of any changes in the system's context to the automation software and facilitates the embedding of the automation software into the overall plant design.

Refinement of logical architecture

During automation software design, the logical viewpoint is refined and enriched with typical automation software artifacts such as continuous function charts and sequential function charts, see [IEC 61131]. Fig. 12-8 shows an example continuous function chart that is part of the seawater pumping and controls beach well 1. Continuous function charts allow the creation of automation software based on predefined function blocks, which are shown as boxes in Fig. 12-8.

Sequential function charts supplement the continuous function charts with the possibility to control sequential or parallel processes which are discrete in time or are event-driven. Fig. 12-9 shows a sequential function chart responsible for controlling the beach wells.

With the refinement and enrichment of the logical viewpoint, the scope of software design is shifted by setting the focus to the next deeper degree of granularity according to the SPES XT modeling framework, see Chapter 3.

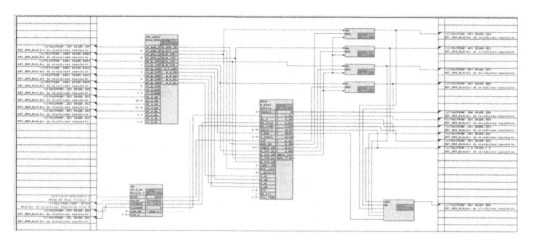

Fig. 12-8   *Continuous function chart (Siemens SIMATIC PCS 7)*

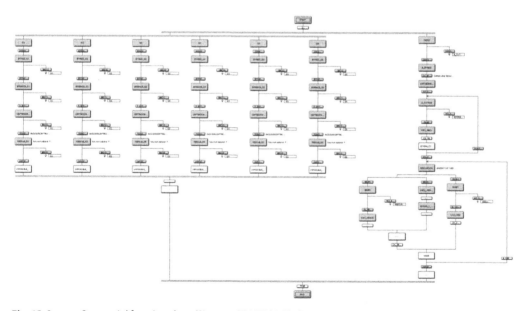

Fig. 12-9   *Sequential function chart (Siemens SIMATIC PCS 7)*

12.4.4 Technical Viewpoint

With regard to the technical viewpoint in automation, software engineering basically means configuring the automation hardware and assigning the logical components of automation software to the

Configuration and assignment

hardware. The automation hardware itself consists of standard hardware components which are listed in hardware component libraries.

During hardware configuration, the automation hardware and infrastructure is defined and configured — for example, which CPUs are used, which bus system and device gets which address, etc. Fig. 12-10 shows an excerpt of a hardware configuration of a desalination plant with two redundant racks with modular I/O stations and actuators which control the valves in the desalination plant. Because the requirements of the runtime behavior of the automation software are usually strict, the user assigns software parts to specific CPUs and defines the update cycle of the automation software execution during operation.

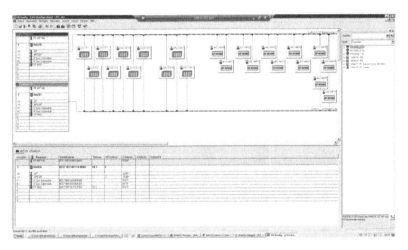

Fig. 12-10   *Hardware configuration (Siemens SIMATIC PCS 7)*

12.5   Implication for Engineering Tools Used Today

Automation engineering tool chain

A typical automation engineering tool chain which is used in a plant engineering process as described in Section 12.2 consists of Siemens COMOS and Siemens SIMATIC PCS 7. Siemens COMOS is an engineering tool for the design, operation, and management of process plants, for example, pharmaceutical or chemical plants. It integrates engineering data of various disciplines and thus provides an integrated view of plant engineering. Siemens COMOS is used in particular as a process engineering tool. It is tightly integrated with Siemens SIMATIC PCS 7 with respect to the automation-relevant data. SIMATIC PCS 7 is a distributed process control system also mainly for process industries, see also [Urbas 2012]. It supports the design of the automation software as well as configuration and integration of the necessary automation hardware.

When looking at the artifacts delivered in the automation do-main today (see Fig. 12-1), we can distinguish between artifacts which concentrate on the automation software engineering — that is, the SUD — and artifacts which address the entire plant. Infor-mation in the later artifacts which is relevant for automation soft-ware engineering has to be associated to the context models of the corresponding viewpoint (see Chapter 4).

Typical artifact types

With respect to the engineering tools, information in the engi-neering artifacts created by Siemens COMOS is mostly associated with the context models of the SPES XT context modeling frame-work (see Chapter 4), while the engineering artifacts created by Siemens SIMATIC PCS 7 mostly address the automation software — that is, the SUD: continuous function charts and sequential func-tion charts are real automation software artifacts within the SUD. We could use the continuous function charts to describe exclusively the functional viewpoint but the continuous function charts will usually still contain technical attributes.

Association to the context models

In today's automation software engineering, the logical view-point is not modeled explicitly. The logical architecture is described implicitly, for example, by continuous function charts and by the arrangement of symbols and use of scripts in the operator pictures (HMI). Today's scattered representation of the logical architecture originates from the different sub-disciplines which are involved in automation software engineering (i.e., hardware engineering, HMI design, and control engineering) and the close collaboration with neighboring disciplines (e.g., mechanical and electrical engineering).

Although today the functional and logical architectures of auto-mation software are not modeled in a strict way, the technical viewpoint is clearly separated by actual process control systems which provide preselected hardware components and allow only valid hardware configurations.

12.6 Summary

Today, the artifacts of automation software design do not distin-guish between functional and logical viewpoints. Moreover, they also incorporate details of the technical viewpoint since the automa-tion software and hardware are closely interlaced. Nevertheless, we can apply the SPES XT modeling framework even when we are using today's engineering tools and complying with strict application and implementation guidelines.

The system's context is very important for automation software engineering and there are also numerous interdependencies between the viewpoints of the SPES XT modeling framework and also in-between. In general, there are a variety of changes — even late in the plant engineering process — which affect the automation software and must be taken into account. In order to keep the effects of such changes under control with minimal efforts, the systematic approach of the SPES XT modeling framework to explicitly document dependencies between the context and the automation software as well as inside the automation software improves engineering efficiency.

All in all, most principles of the SPES XT modeling framework are already applied in the automation domain. However, they are not as transparent and clearly visible to a user as they should be. In particular, the dependencies between the different artifacts concerning the SUD as well as the system's context are often not explicitly documented but naming conventions are used by the automation software developers and the neighboring disciplines to express these dependencies.

In the automation domain, there are already approaches which address the challenges that arise due to critical dependencies in mechatronic systems. However, these approaches are often implicit and rely on a common understanding by all stakeholders. All in all, the SPES XT modeling framework serves as a guide to more formalization of automation engineering.

12.7 References

[ANSI/ISA-88.01-1995] Instrument Society of America: ANSI/ISA-88.01-1995: Batch Control, Part 1: Models and Terminology, 1995.

[Campetelli et al. 2015] A. Campetelli, M. Junker, V. Koutsoumpas, X. Zhu, B. Boehm, M. Davidich, J. C. Wehrstedt: A Model-Based Approach to Formal Verification in Early Development Phases: A Desalination Plant Case Study. Fünfter Workshop zur Zukunft der Entwicklung softwareintensiver eingebetteter Systeme (ENVISION 2015), 2015.

[Fay 2009] A. Fay: Effizientes Engineering komplexer Automatisierungssysteme. In: E. Schnieder, T. Ständer (Eds.): Wird der Verkehr automatisch sicherer?: 20 Jahre - vom IfRA zum iVA - Jubiläumskolloquium; Beschreibungsmittel, Methoden und Werkzeuge des integrierten Systementwurfs zur Fahrzeug- und Verkehrsautomatisierung; 04. September 2009 in Braunschweig. Braunschweig: iVA, S. 43–60, 2009.

[Holm et al. 2013] T. Holm, S. Schröck, A. Fay, T. Jäger, U. Löwen: Engineering von "Mechatronik und Software" in automatisierten Anlagen: Anforderungen und Stand der Technik. In: S. Wagner, H. Lichter: Software Engineering 2013 Workshopband, Aachen, Germany, February 26 – March 1, 2013, Proceedings Lecture Notes in Informatics. Gesellschaft für Informatik, Bonn, 2013, pp. 261-272.

[IEC 61131] IEC: IEC 61131-3:2013 – Programmable controllers - Part 3: Programming languages. Edition 3.0, 2013.

[Urbas 2012] L. Urbas: Process Control Systems Engineering. Oldenbourg Industrieverlag, Munich, 2012.

Torsten Bandyszak
Philipp Diebold
André Heuer
Thomas Kuhn
Antonio Vetrò
Thorsten Weyer

13

Technology Transfer Concepts

In software engineering, transferring innovative concepts, techniques and methods into the practice of existing organizations is an expensive and complex task. This chapter gives an overview on the transfer of the SPES XT modeling framework to different organizations. The focus of this chapter is threefold: first, it addresses the characterization of technology transfer in applied software engineering research. Second, it focuses on the concept of guidelines that we have specifically developed for technology transfer. Third, this chapter presents the artifact quality assessment framework, a tool developed to assess the successful introduction of results from software engineering research allowing us to measure the quality of engineering artifacts created by introduced concepts, techniques, and methods.

© Springer International Publishing AG 2016
K. Pohl et al. (eds.), *Advanced Model-Based Engineering of Embedded Systems*,
DOI 10.1007/978-3-319-48003-9_13

13.1   Introduction

Importance of technology transfer

Technology transfer subsumes activities which enable potential users to adopt and apply solutions from research, such as concepts, techniques, methods, or tools, in practice (cf. [Teece 1977]). In software engineering in particular, technology transfer is essential since past evidence shows that technology developed in research projects can take up 15-20 years to achieve widespread adoption [Redwine and Riddle 1985]. Hence, appropriate technology transfer concepts have to be developed and implemented to support a light-weight and seamless integration of state-of-the-art techniques into practice. In this regard, the positive evaluation results in industrial case studies presented in Chapter 12 indicate the feasibility of concepts and techniques of the SPES XT modeling framework. However, this does not automatically imply that this technology is ready to be transferred into a real industrial setting as it is. Therefore, we have analysed and characterized how technology transfer occurred in the SPES XT project, understanding its limitations, how to improve it, and trying to facilitate it.

Technology transfer in SPES XT

Given the importance of evaluating the technology transfer of developed technological solutions [Pressman 1988], in this chapter we present central concepts used for assessing the technology transfer of the SPES XT modeling framework into industry. In doing so, we present the patterns observed, with a main focus on the benefits and barriers of the techniques developed and on the media used to transfer them to practice (Section 13.2). Regarding the latter, in Section 13.3 we present the guideline documents which were developed specifically for supporting technology transfer in the SPES XT project. Finally, in Section 13.4, we describe the *artifact quality assessment framework* (AQAF) which aims at supporting the comparison of the quality of different engineering artifacts.

13.2   Technology Transfer in SPES XT

Goal of technology transfer study

Based on the execution of two surveys [Diebold and Vetrò 2014] [Diebold et al. 2015] and on a linear model of technology transfer widely used in literature [Teece 1977], our goal was to *characterize the current state of practice in technology transfer in SPES XT, the transfer media used, and to identify directions for improvements*. Here we report the patterns observed.

An important and practical implication of this finding was to understand which techniques developed in SPES XT could be adopted as they are and which encounter the typical barriers to the adoption of model-driven development (MDD) (see [Torchiano et al. 2013]), meaning that strong customization and post-project-specific support is required. In general, the barriers to the adoption of certain techniques are mainly the lack of supporting tools, the lack of proper competencies, and the high effort required to introduce the techniques (the latter might be a consequence of the first two). In support of these barriers, we found that the two most frequently cited benefits of the existing model-based engineering approaches are the tool support and the ease of use. On the benefits side, the most frequently cited expected incentives for investing in the adoption of the SPES XT techniques are design support, better quality of software, and documentation.

SPES XT techniques need customization to fit industrial contexts

Beyond a lack of a well-defined process, some respondents also reported informal ways of conducting a transfer: we know from literature that informal transfers are more probable in the case of spatial proximity. In addition, the answers revealed that technology transfer is not perceived as a continuous process, especially by the transferees. Future effort can be devoted to the creation of an ad-hoc process for the transfer to practice.

Technology transfer takes place without an explicit process

Although there is abundant evidence that external knowledge has a positive impact on product development and innovation [Katila et al. 2002], the main trigger for innovation in the SPES XT project occurred from within the boundaries of the organizations. Future effort should be devoted to reinforcing the communication flows between the partners.

Industry organizations gain new knowledge mainly on their own

This finding was surprising because of the guidelines which were the main transfer medium in SPES XT (see Section 13.3). Nonetheless, a detailed investigation divided into the organizational types — that is, academia vs. industry — showed that guidelines as a transfer medium were among the media most commonly used by industry partners. In general, we observed that unsuitability and low resources were the main motivation for low attendance at events, while lack of awareness of existence was the main reason for not using artifact-oriented media. Regarding the achievement of purpose when using a medium, we can report that the most successful media were the SPES XT website, the mailing lists, the publications, and the SPES 2020 conference in 2012. Technical artifacts (scenarios, building blocks, guidelines), the SPES wiki, SPES results for

Human-intensive media are favored

academic courses, surveys, and individual dissemination activities had a positive success rate.

Heterogeneous drivers for transfer

We observed that the motivation for technology transfer differs heavily between industrial and academic organizations, as expected. Technology transfer in industry is to a large extent motivated by economic and strategic objectives. In contrast, in the academic area technology transfer is driven by more personal reasons — for example, intellectual growth.

Transferor and transferee are not clearly distinguishable

We believe that the traditional way of modeling technology transfer is not precise enough to capture the complexity of the interactions involved in the transfer. In fact, many of our respondents identified themselves on both sides — as transferors *and* as transferees. Based on this observation, we reworked the original definition of technology transfer given by [Teece 1977] to produce a new one [Diebold et al. 2015] which should correspond more closely to what we expect to hold true for software engineering projects:

A new definition for technology transfer

> *Technology transfer is the process of sharing or developing a technology object between two or more actors via one or more media so that the technology recipient sustainably adopts the object in the recipient's context in order to evidently achieve a specific purpose.*

13.3 Guideline Concepts

Using a guideline concept has been proven as a comprehensible approach for transferring technology (cf. [Vieira et al. 2012]). The results from the previously presented study in Section 13.2 support this idea of using guidelines as transfer media, especially for the industrial partners.

This section presents generic recommendations which support the creation of domain-specific guidelines for the automation, avionics, and automotive industries. These recommendations define a generic guideline concept that addresses key requirements for the applicability of research results collected from industry and literature.

Process of creating the recommendations

The guideline concept is developed in an iterative way which allows the integration of feedback from the application in industry in order to incrementally increase its maturity. Further details on the guideline concept and the corresponding recommendations can be found in [Heuer et al. 2014].

13.3.1 Key Requirements for Guidelines

Requirements for guidelines cover, amongst other things, the structure of the document with respect to technology introduction phases and activities and the use of examples to illustrate the concepts. In addition, general considerations about technology transfer also constitute valuable sources for requirements, since guidelines aim at facilitating and supporting change processes. Goal-orientation — for example, benefit analysis — and addressing the management are key factors for successful technology transfer and should be reflected in guidelines. An excerpt of the resulting set of requirements from specific application domains is shown in Tab. 13-1.

Tab. 13-1   *Representative subset of guideline requirements (cf. [Heuer et al. 2014])*

Req.	Title	Domain(s)
R-1	A guideline should consider the organizational context of a given technology.	automation, automotive
R-2	A guideline should emphasize the goals and benefits of the technology that is to be introduced.	automation, automotive
R-5	A guideline should enable an estimation of the effort for technology introduction.	automation
R-16	A guideline should be independent of specific tools.	avionics

13.3.2 Recommendations for Writing Structured Guidelines

Based on the requirements, a reference structure for guidelines was designed. Although all requirements should be addressed, not all of them are directly reflected in chapter headings in the structure. For instance, R-16 from Tab. 13-1 is addressed by a guideline document as a whole. The generic reference structure of guidelines is mainly driven by two aspects: (1) the use case of the guideline, and (2) the target audience of the guideline.

First, the audience of the desired guideline has to be defined, (i.e., whether it is used within a company or across several companies). Company-internal guidelines can be more specific and tailored to the company's needs. In contrast, cross-company guidelines should not mention specific contact persons or departments of a company and might stay at a more abstract level. In each guideline, a set of reader groups, in terms of roles within a company, is addressed by the guideline. The following groups of guideline users are addressed:

❑ *Decision makers* decide whether the proposed technology should be introduced in the company/department or not.

Different reader groups for guidelines

❏ *Coordinators* are responsible for introducing a new technology after a positive decision by the decision makers.
❏ *End users* are those employees who have to apply a new technology within the company.

Tab. 13-2 shows the table of contents that constitutes the reference structure. Most chapters in this structure reflect the majority of requirements, as, for example, requirement R-2 from Tab. 13-1: this requirement is addressed by including a chapter that describes the purpose, that is, the goals and benefits of the technology that is to be introduced.

Tab. 13-2   *Reference structure of guidelines, including chapter titles*

Chapter	Target audience		
	Decision makers	Co-ordinators	End users
Preamble	✓	✓	✓
1. **Introduction**	✓	✓	✓
1.1. Classification			
1.2. Motivation and problem description			
1.3. Purpose of the technology			
2. **Management**			
2.1. Field of application	✓	✓	-
2.2. Risks and challenges			
2.3. Effort for introduction and usage			
2.4. Pilot project			
3. **Technology context**			
3.1. Organizational context	-	✓	-
3.2. Process context			
3.3. Technical context			
3.4. Social context			
4. **Technology application**			
4.1. Preconditions	-	-	✓
4.2. Methodological process			
4.3. Postconditions			
5. **Glossary**	✓	✓	✓
6. **References**	✓	✓	✓

Guiding questions support authoring of guidelines

Within this reference structure, the specific guideline chapters are augmented by questions. Those guiding questions should support the author for each of the chapters. The concept of guiding questions is borrowed from education and didactics. Here, these questions should enable authors to properly design a guideline, achieving their goals to successfully transfer research results to industry. The guiding questions address different aspects concerning the required information, structures to be created, decisions to be made, and expected effects on the part of the organization.

13.4 Artifact Quality Assessment Framework

Coverage of viewpoints and granularity layers in model-driven development (MDD) processes requires the use of different models that cover, for example, structural and behavioral aspects of the embedded system under development. Ensuring the quality of these models requires the mapping of abstract quality attributes such as maintainability, extensibility, and traceability to specific rules that conserve knowledge regarding model quality. While these rules exist for defined contexts, a transfer is challenging, however, due to a lack of formalism and expressiveness. Standards only provide high-level quality goals that cannot be easily mapped to specific metrics. This is also due to the fact that specific quality measurements depend on model elements and metamodels; they cannot be distributed as plug-ins because they need to be adapted.

Ensuring quality of models

Knowledge transfer therefore has to address this problem on a different level. Rules have to be specified on a more abstract level: multiple modeling languages are used for embedded systems specification. Standard modeling languages (such as UML and SysML) are too generic to cover all specialized aspects of embedded systems modeling — they have to be adapted and extended for project and domain contexts. Conservation and transfer of quality knowledge in reuseable quality rules must be sufficiently abstract to allow reuse in the context of different modeling languages and language profiles, as well as flexible tool chains that may be adapted with respect to modeling languages and quality indicators.

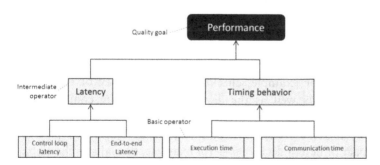

Fig. 13-1   *Quality tree example for performance*

Quality trees provide a high-level and implementation-independent

Quality trees

description of quality attributes and therefore enable this transfer. They define data dependencies and operations for quality measurements. Leaves of quality trees define the collection of quality metrics from models. The example in Fig. 13-1 illustrates a quality tree that describes the calculation of expected performance based on base metrics that are collected from input models. Basic operators extract basic metrics from models. Intermediate operators calculate derived metrics — for example, expected latency and timing behavior. Quality attributes are calculated based on the outputs of these operators. Each operator can be specified using a data flow graph that defines its algorithm.

To support the setup of tailored quality measurement programs, existing and proven quality metrics have been collected from literature and project partners.

Evaluation of quality indicators

One tool that enables the evaluation of models according to quality indicators is INProVE [Basili and Rombach 1988]. The INProVE (Indicator-based Non-functional Property-oriented eValuation and Evolution of software design models) tool supports the implementation of tailored quality measurement programs that use an adapted goal question metric (GQM) [Kemmann et al. 2010] approach. Its overall structure is illustrated in the figure below. INProVE uses model adapters to connect to design models — this enables the integration of new model types that were initially not supported. Model adapters need to convert input models into Eclipse Modeling Framework (EMF) models. As an option, model adapters may target an existing metamodel, for example, EMF-UML2. This enables the use of existing INProVE operators for these metamodels with the newly integrated models. INProVE already defines a set of operations for the implementation of quality indicators, including basic operations such as mathematical and set operations, as well as more advanced operations including fuzzy logic. Quality indicators are defined in a graphical data flow language that resembles the structure of quality trees. These support the development of basic metrics that directly relate to model elements — for example, checking the consistency of two or more models — as well as the development of more complex derived metrics that are based on the basic metrics and combine them using fuzzy logic, for example. This enables developers to predict the impact of design decisions — for example, to evaluate the change in unit cost depending on deployment decisions and the derived need of hardware devices. Results are stored in a database. Standard

tools (e.g., Matlab, Excel) may connect to this database and visualize the results depending on project needs.

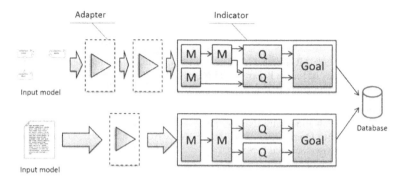

Fig. 13-2   *Concept of the INProVE tool*

13.5 Summary

The technology transfer concepts presented in this chapter can serve
as a foundation for more advanced and elaborate concepts. Method
specialists (either company-specific or independent), for example,
could train employees of the companies in workshops for specific
methods. Based on workshops, guidelines can act as a reference. In
addition, existing guidelines, and thus the guideline structure, could
be extended by cheat sheets. The guidelines itself are specifically
tailored for specific domains. However, as the SPES XT modeling
framework could also be used in other domains, the technology
transfer concepts, especially the guidelines, may need to be adapted
to deal with further domain-specific requirements.

13.6 References

[Basili and Rombach 1988] V. R. Basili, H. D. Rombach: The TAME project. Towards improvement-oriented software environments. In: IEEE Transactions on
   Software Engineering. Bd. 14, Nr. 6, 1988, pp. 758–773.

[Diebold and Vetrò 2014] P. Diebold, A. Vetrò: Bridging the gap: SE technology
   transfer into practice: study design and preliminary results. In: Proceedings of the
   8th ACM/IEEE International Symposium on Empirical Software Engineering and
   Measurement (ESEM '14), ACM, New York, NY, USA, Article 52, 2014.

[Diebold et al. 2015] P. Diebold, A. Vetrò, D.M. Férnández, An Exploratory Study
   on Technology Transfer in Software Engineering. In: Proceedings of the 9th
   ACM/IEEE International Symposium on Empirical Software Engineering and
   Measurement (ESEM '15), ACM, New York, NY, USA, 2015.

[Heuer et al. 2014] A. Heuer, P. Diebold, T. Bandyszak: Supporting Technology Transfer by Providing Recommendations for Writing Structured Guidelines. In: K. Schmid, W. Böhm, R. Heinrich, A. Herrmann, A. Hoffmann, D. Landes, M. Konersmann, T. Ruhroth, O. Sander, V. Stolz, B. T. Widemann, R. Weißbach: Software Engineering Workshops 2014 (SE-WS 2014) - Gemeinsamer Tagungsband der Workshops der Tagung Software Engineering 2014, Vol. 1129, CEUR-WS.org, 2014, pp. 47-56.

[Katila et al. 2002] R. Katila, G. Ahuja. Something Old, Something New: A Longitudinal Study of Search Behavior and New Product Introduction. The Academy of Management Journal, Vol. 45 No. 6, 2002, pp. 1183–1194.

[Kemmann et al. 2010] S. Kemmann, T. Kuhn, M. Trapp: Extensible and Automated Model-Evaluations with INProVE. In: System Analysis and Modeling: About Models: 6th International Workshop, SAM 2010, Oslo, Norway, October 4-5, 2010, Revised Selected Papers

[Pressman 1988] R. S. Pressman: Making Software Engineering Happen – A Guide for Instituting the Technology. Prentice Hall, Englewood Cliffs, New Jersey, 1988.

[Redwine and Riddle 1985] S. T. Redwine, W. E. Riddle: Software Technology Maturation. Proceedings of the 8th International Conference on Software Engineering. IEEE Computer Society Press, Los Alamitos, California, 1985, pp. 189-200.

[Teece 1977] D. J. Teece: Technology transfer by multinational firms: The resource cost of transferring technological know-how. In: Economics Journal, Vol. 87. No. 346, 1977, pp. 242-261.

[Torchiano et al. 2013] M. Torchiano, F. Tomassetti, F. Ricca, A. Tiso, G. Reggio: Relevance, Benefits, and Problems of Software Modelling and Model-Driven Techniques - A Survey in the Italian Industry. In: Journal of Systems and Software, Vol. 86 No. 8, 2013, pp. 2110-2126.

[Vieira et al. 2012] E. R. Vieira, C. Alves, L. Duboc: Creativity Patterns Guide – Support for the Application of Creativity Techniques in Requirements Engineering. In: M. Winckler, P. Forbrig, R. Bernhaupt (Eds.): Human-Centered Software Engineering, Proceedings of the 4th International Conference on HCSE, Toulouse, 2012, LNCS, Vol. 7623. Springer, Berlin Heidelberg, 2012, pp. 283-290.

Karsten Albers
Ömer Gürsoy
Stefan Henkler
Michael Schulze
Bastian Tenbergen
Axel Terfloth
Raphael Weber

14

The SPES XT Tool Platform

The SPES XT modeling framework allows seamless, model-based development of complex embedded systems. While the SPES XT modeling framework is defined on a conceptual level, a technical realization is required for industrial success. More specifically, an extensible and adaptable tool platform is required which supports interoperability, reuse, and traceability of methods and tools and provides a common view of artifacts. In this chapter, we introduce the SPES XT tool platform, which addresses these requirements by integrating the SPES XT modeling framework with established concepts.

© Springer International Publishing AG 2016   251
K. Pohl et al. (eds.), *Advanced Model-Based Engineering of Embedded Systems*,
DOI 10.1007/978-3-319-48003-9_14

14.1 Introduction

Our objective is to define an SPES XT tool platform, thus providing an envelope for complex system design and covering a large set of formally specified concepts enabling seamless, model-based development of embedded systems. The SPES XT tool platform places particular emphasis on the integration with existing development and analysis tools while building on a common methodology defined by the SPES XT modeling framework and interoperability concepts.

Integration of tools, services, and methods

The SPES XT tool platform comprises three main concepts: on the one hand, a common methodology has been defined in order to establish a semantic basis for tool integration. This supports the integration of tools with regard to organizational differences in the development process and allows the substitution of particular tools in a tool chain with another tool supplied by a different vendor. On the other hand, an interoperability concept describes the meaning of data and services required or provided by tools. Based on this interoperability concept, the common methodology allows us to relate these data and services and build tool chains on a conceptual level.

Relevant integration aspects

While a common methodology is of paramount importance for tool integration, it is also important to define how tools can be integrated and what interoperability concepts are used. While the common methodology defines *what* information is exchanged between tools, the interoperability concept defines *how* this information is exchanged.

Because interoperability and data integration largely depend on organizational circumstances, the SPES XT tool platform consists of three concept layers in order to abstract from tool-specific or development process-specific considerations. Section 14.2 illustrates the interoperability concepts. Section 14.3 then describes the SPES XT tool platform.

14.2 Interoperability and Tool Integration Concepts

As outlined above, there are several levels of interoperability of methods and tools. These differences in interoperability can be addressed using a variety of established techniques. For example, interoperability at tool level is supported by making use of what is known as a pivot model (see Section 14.2.2). Furthermore, *Open*

Services for Lifecycle Collaboration (OSLC) is an open standard for tool interoperability (see Section 14.2.3). Similarly, more generic frameworks for interoperability have been developed by projects in the past (e.g., CESAR; http://www.cesarproject.eu) and have been successfully extended (e.g., CRYSTAL; http://www.crystal-artemis.eu). In order to draw on the benefits of established technologies, the SPES XT tool platform was designed to conform to the CESAR interoperability platform to enable tool integration in a broader community.

14.2.1 Integration Levels and Interoperability Layers

There are many strategies for tool integration which we categorize into three integration levels. Basic approaches address the integration of data and allow seamless integration of different data fragments and coordination of tool access to this data. The operation-level integration addresses the collaboration of tools. In this case, tools have to interact with one another. Simple invocation interfaces or more complex APIs or services that are provided by tools may be used to achieve this. On the next level, interaction aspects are addressed in the sense that the integrated tools collaborate in providing an integrated user interface or on an organizational level, integrate into a common workflow. Fig. 14-1 gives an overview of these levels of integration.

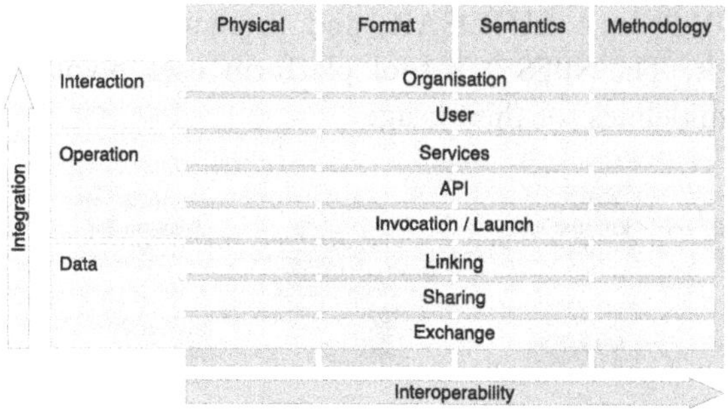

Fig. 14-1   *Tool integration strategies and interoperability layers*

Any kind of tool integration requires tool interoperability. Interoperability is crosscutting to integration and different interoperability layers can be identified, as shown in Fig. 14-1. The basic physical layer defines the medium that is used for interoperation, such as the use of files, web services, APIs, or HTML5 as user interface technology. The format layer defines the specific syntax for data or operational signatures. The semantic layer defines everything that is necessary to establish a common understanding of the meaning of the definitions on the format layer. Finally, the methodology layer defines the purpose of interoperation and the role of each tool. This layer addresses the context of the tool platform and is typically prescribed by the application domain.

14.2.2 Pivot Model

The use of pivot models is an approach to tool interoperability and integration. We define a pivot model as follows:

Definition

A pivot model is the central representation of model artifacts on which tools interoperate.

A pivot model requires a well-defined metamodel and specific representation which allows interoperability on a physical, syntactical, and semantic layer and thus enables data-level integration. A conceptual overview of the pivot model is shown in Fig. 14-2.

Standardization of model data

A pivot model implies standardization of model data and allows the integration of alternative model formats by introducing transformations to and from the pivot model representation. It helps to avoid point-to-point integrations between any two model representations and thus reduces integration efforts. As a consequence, all tools directly or indirectly connected to the pivot model may utilize many model sources. Examples for pivot models are Franca-IDL [Birken 2015] and AUTOSAR, and in many cases, UML-based metamodels. The SPES XT tool platform uses pivot models and provides guidelines for their usage.

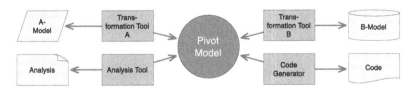

Fig. 14-2   *Pivot model as a concept for integrating methods and tools*

14.2.3 Linked Data and Services

Open Services for Lifecycle Collaboration

The goal of the *Open Services for Lifecycle Collaboration* (OSLC) core specification [Johnson and Speicher 2013] is to provide solutions for tool integration. The basic idea is that tools should expose their data and functionality by means of web services. Based on the principles of *linked data*, RESTful protocols can be used to link data provided by different tools in a similar manner to creating hypertext links between webpages. Tool integration is hence achieved by means of web services and is thus transparent to particular programming languages or platforms and even allows the integration of tools running on different computers. There are various OSLC specifications (which extend the core specification) detailing which data is expected to be defined for certain notions. For example, the OSLC RM specification [Johnson and Speicher 2013] declares that a requirement should always have a rationale.

The OSLC specifications are based on existing established standards: HyperText Transfer Protocol (HTTP) and Resource Description Framework (RDF). Each artifact provided by a tool is exposed as an HTTP resource in OSLC that is manipulated by means of HTTP commands such as GET, PUT, POST, and DELETE. The RDF standard is used to build representations of these HTTP resources. The OSLC core specification then defines the protocol which regulates the way OSLC-conformant tools interface with one another through HTTP commands. In addition, the OSLC specification defines a number of resource types and their properties which allow us to discover the data and functionalities provided by tools.

OSLC specifications are based on standards

14.3 Defining the SPES XT Tool Platform

As outlined in Section 1.1, the SPES XT tool platform has to ensure interoperability, reusability, and traceability of methods and tools, provide a common view of artifacts, and be extensible and adaptable such that additional methods and tools can be easily integrated with regard to specific organizational needs. In the following, we detail the SPES XT tool platform and explain its three conceptual layers. An overview over the SPES XT tool platform can be seen in Fig. 14-3.

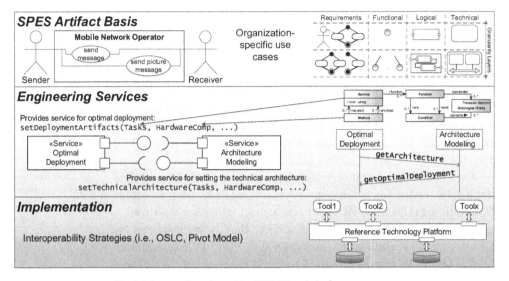

Fig. 14-3   *Overview of the SPES XT tool platform*

Overview of the SPES XT tool platform

On the uppermost layer, we introduce the technical definition of the SPES XT artifacts (SPES Artifact Basis) based on the SPES XT modeling framework. The SPES XT tool platform uses the dependencies between viewpoints and artifacts to allow traceability between SPES artifacts. The *Engineering Services* layer defines the services required to realize the *use cases* of the different SPES XT modeling framework applications (see Part II) in a tool-independent manner. In the *Implementation* layer, platform-specific realizations of the engineering services are defined in accordance with CESAR. These three layers are explained in more detail in the following sections.

14.3.1 First Layer: Common Artifact Basis

The SPES XT modeling framework combines viewpoints and granularity layers which form an engineering space as illustrated in Chapter 3. Because the SPES XT modeling framework is generic in the sense that it can be adapted for a variety of development processes (see Chapters 3 and 5), it does not dictate a specific order in which artifacts are to be developed and documented. For this purpose, well-defined dependencies between viewpoints across multiple granularity layers are used to decide the specific semantics of one artifact in one viewpoint in one granularity layer from the perspective of another artifact. This gives rise to a multitude of possible *engineering paths* through the engineering space, that is, a develop-

ment process-specific and organization-specific order in which artifacts are developed and in which the system is decomposed into
more detailed subsystems. Each of these subsystems can be regarded
as a separate system under development for which the development
process starts anew by means of the same structure of SPES artifacts. The aforementioned engineering paths thereby serve as a basis
for defining specific methods which support various development
steps/decisions. Tool support for these development steps may be
defined as services in the service layer which is described in the next
section.

14.3.2 Second Layer: Service Platform

As outlined in Section 14.2.1, an engineering service is a conceptual
representation of a repeatable development activity that has a specified outcome. For example, this may include artifact validation
tasks and automated artifact transformations or consistency checks.
Furthermore, an engineering service is self-contained and may sub-
sume other engineering services, similar to service-oriented architectures [Biehl et al. 2012]. In this sense, the engineering services in the
SPES XT tool platform are akin to the methodological building
blocks described in Chapters 6 through 11, yet have a more technical connotation. In the SPES XT tool platform, it is assumed that
each methodological building block is supported by some technology assisting it.

In the SPES XT tool platform, the following types of engineering
services can be differentiated:

Types of engineering services

❑ *Access and manipulation services*, which coordinate basic operations on artifacts, such as creation, reading, updating, deletion
  (CRUD).
❑ *Transformation services*, which allow the manual or automated
  transformation of one or more artifacts into other artifacts (e.g.,
  automatic generation of partial function networks from scenar-
  io-based requirements, see Chapter 9).
❑ *Analysis services*, which allow the validation and verification of
  one or more artifacts and produce analysis results or, in some
  cases, counterexamples (e.g., timing analysis of executable specifications in Chapter 11).

All these services operate on artifacts and address the integration on
data and operation levels. The service platform advises that these
services are stateless, which means that all information necessary to

apply the service should be part of the service's interface. On the next higher integration level, user interface services can be defined (see Section 14.2.1). Due to the interactive nature of many user interface-bearing tools, user interface services must capture the state of interaction with the user and are thus stateful.

Service metamodel   Services on the service platform must be defined based on a service metamodel in order to generate service function implementation skeletons which may then be filled with behavioral code. This is subject to the third layer described in Section 14.3.3. The service metamodel of the SPES XT tool platform is shown in Fig. 14-4.

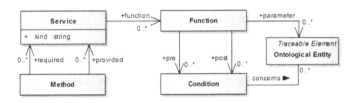

Fig. 14-4   *SPES XT tool platform's service metamodel*

In general, a method is described by attributes for the general description of the method (e.g., what the method does, why this is useful, etc.) and the method name. A method has required and provided services which have simple service functions which are typically small parts of the service that do not need user interaction. The service functions are defined on the artifacts of the SPES XT modeling framework. Furthermore, a service function is described by preconditions and postconditions in the form of textual statements or sequence diagrams. For an example of such a service description, see Section 7.3.

Basic services   Based on the concepts of the service platform layer, the SPES XT tool platform defines basic services. The common property of basic services is their crosscutting usage for the different methodological building blocks in Chapters 6 through 11. For example, for embedded systems with hard real-time and safety characteristics, tracing may be such a basic service. For another example, from the perspective of variability management and structured reuse, the handling of different product variants may be a basic service. In the following, these basic crosscutting services are described in more detail and an example is given.

Trace Management Services

Managing relationships between different engineering artifacts is a crosscutting task that provides the foundation for traceability. Traceability — especially in the sense of requirements traceability — helps to ensure an overall consistency of engineering artifacts and to control change impact. More details on traceability between SPES viewpoints, granularity layers, and artifacts can be found in Chapter 3. Furthermore, an approach for validating trace links between artifacts was outlined in Section 11.3.2. As traceability covers all artifacts within the development scope of a product or project, it makes sense to provide a common infrastructure for managing trace data. Fig. 14-5 gives an overview of the different trace management services defined by the SPES XT tool platform.

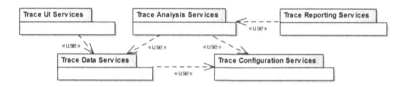

Fig. 14-5   *Trace management services pertaining to all SPES XT artifacts*

The trace management services follow the different service categories and integration level and support interoperability up to the methodology layer shown in Fig. 14-1. The service definitions build upon a trace data model that serves as a pivot model for trace data, as shown in Section 14.2.2. The trace management services and data model are discussed in detail in Chapter 11.

Variant Management Services

Managing variation points, variations, and dependencies between them is a crosscutting task and this crosscutting nature is also reflected through the introduction of the variability perspective, which spans all SPES viewpoints and granularity layers. A detailed description of the variability perspective is given in Section 11.2.1. The general motivation for variant management is to support the systematic reuse of artifacts and enable the building of a product line in contrast to just a single product.

The specified variant management is the foundation for managing a product line. For example, the variant management services deliver functionality for connecting an artifact with features, defining an existence condition, and storing that information. Service functions for analyzing change impacts on variants and for transforming product line artifacts into product-specific/variant-specific artifacts are provided. The formalized and modeled variability knowledge is used to ensure that only valid variants can be generated.

The variant management services, the corresponding data model, and the specification of the variability exchange language are discussed in detail in [Schulze and Hellebrand 2015].

Example Service Description

An example of a service description is the optimization based on design space exploration (DSE) which is part of the technical viewpoint. As input, the service receives the fixed logical architecture and the destination hardware architecture (which can include some degree of freedom with regard to the number and type of hardware elements). The output is, if applicable, the selected technical architecture, the mapping of the logical architecture on the technical architecture, and the values of the scheduling parameters (such as priorities) required for a reasonable timing which also fulfills the timing requirements.

The services for the DSE consist of a whole set of different methods which can be selected and combined depending on the specific problem. For example, one method can generate distributions of the logical architecture to the technical architecture using other methods in the loop which explore and select the appropriate scheduling parameters for the given mapping. The selected parameters are required to rate the different candidates for the mapping. See the method description in Chapter 9 for more details.

14.3.3 Third Layer: Implementation Platform

The third layer is the implementation platform layer which subsumes implementation-centered interoperability concepts. As an implementation basis, it is possible to use OSLC at the most generic level, as discussed in Section 14.2.3. However, there are other ways to implement tool interoperability based on the service platform layer. For example, the plug-in mechanism in the Eclipse Platform [Gamma and Beck 2003] allows interoperability programmers to link services (plug-ins) to exchange information without an OSLC interface. It might be easier to use Eclipse mechanisms already provided to make the plug-ins interoperable.

Fig. 14-6 shows an example of how different components can be linked in order to interoperate with each other. Outside Eclipse, *Tool* and *Service* can be linked to the Eclipse Platform via an OSLC interface. Within the Eclipse Platform, plug-ins may also be linked to each other using Eclipse interfaces which conform to the SPES service platform.

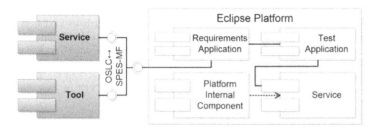

Fig. 14-6   *Tool integration example*

14.4 Summary

In this chapter, we introduced a technical realization of the SPES XT modeling framework (see Chapter 14.2.1) by supporting end-user requirements such as interoperability, reuse, and traceability. To benefit from established technologies, the SPES XT tool platform conforms to interoperability specifications of other projects (e.g., CESAR and its successors) to enable tool integration in a broader community. The SPES XT tool platform thus closes the gap between the seamless modeling approach of the SPES XT modeling framework and the interoperability technologies at data level developed by other projects. Compared to other interoperability approaches, tools integrated with the SPES XT tool platform also have a common understanding of the artifacts exchanged, which is vitally important to support a seamless, model-based design approach. Hence, crosscutting tool integrations (e.g., between the CESAR RTP and the SPES XT tool platform) support only interoperability at data level and not at artifact level.

14.5 References

[Biehl et al. 2012] M. Biehl, J. El-Khoury, F. Loiret, M. Torngren: On the Modeling and Generation of Service-Oriented Tool Chains. In: Journal of Software and Systems Modeling (SoSyM). Vol. 13, No. 2, 2012, pp. 461-480.

[Birken 2015] K. Birken: Franca User Guide. https://franca.eclipselabs.org.codespot.com/files/FrancaUserGuide-0.3.0.pdf. (Accessed October 6, 2015).

[Gamma and Beck 2003] E. Gamma, K. Beck: Contributing to Eclipse: Principles, Patterns, and Plug-Ins. Addison Wesley, 2003.

[Johnson and Speicher 2013] D. Johnson, S. Speicher: Open Services for Lifecycle Collaboration - Core Specification Version 2.0. http://open-services.net/bin/view/Main/OslcCoreSpecification. (Accessed October 6, 2015).

[Schulze and Hellebrand 2015] M. Schulze, R. Hellebrand: Variability Exchange Language – A Generic Exchange Format for Variability Data. In: W. Zimmermann, W. Böhm, C. Grelck, R. Heinrich, R. Jung, M. Konersmann, A. Schlaefer, E. Schmieders, S. Schupp, B. T. Widemann, T. Weyer (Eds.): Software Engineering Workshops 2015 (SE-WS 2015) - Gemeinsamer Tagungsband der Workshops der Tagung Software Engineering 2015, Vol. 1337, CEUR-WS.org, 2015, pp. 71-80.

Philipp Diebold
Marian Daun
André Heuer
Andreas Jedlitschka

15

Evaluation of the SPES XT
Modeling Framework

In software engineering, empirical evaluations play a major role in discovering the advantages and disadvantages of newly developed methods, techniques, and tools. In addition, empirical studies are used to validate the initial rationale behind the development of new approaches. Furthermore, empirical evidence, originating from different kinds of systematically conducted studies, is crucial to support the transfer and application of new methods in industrial settings. This chapter gives an overview of the evaluation actions in the context of the SPES XT modeling framework.

© Springer International Publishing AG 2016
K. Pohl et al. (eds.), *Advanced Model-Based Engineering of Embedded Systems*,
DOI 10.1007/978-3-319-48003-9_15

15.1 Introduction

Objectives of the SPES XT modeling framework
The SPES XT modeling framework targets the structured, model-based engineering of embedded systems. While certain problem-specific approaches are frequently proposed in literature, these approaches lack seamless integration into a consistent engineering framework. Hence, the SPES XT modeling framework aims to support continuous engineering by providing general methods for solving individual problems and allowing reuse and technology transfer across different development units, companies, and even application domains. The SPES XT modeling framework can be applied to various application domains.

Need for empirical evidence
It is very important to provide empirical evidence for the general applicability and usefulness (e.g., in terms of effectiveness or efficiency) of new approaches and to demonstrate how they help to solve industry problems. If such evidence were not provided, the introduction of newly developed methods, technologies, and tools in industrial engineering processes would be hampered. In particular, bear in mind that introducing unsound or inapplicable approaches or approaches with limited usefulness might result in the loss of the company's reputation or monetary resources, and may — in the worst case — lead to complete failure of projects or the company itself.

Evaluation of the SPES XT modeling framework
To satisfy industry's need for empirical evidence regarding the applicability and usefulness of the SPES XT modeling framework, the framework has been evaluated intensively using different study types and by applying the framework to different application domains. In general, this overall evaluation approach provides evidence that the SPES XT modeling framework can be applied to industrial engineering challenges regardless of the specific application domains and can be seen as an appropriate model-based engineering approach for various problem situations.

Study types
It is agreed that one single empirical study cannot fulfill the need for generalizable evidence on its own. Therefore, a combination of several studies is necessary to ensure generalizability [Basili 1992]. For the evaluation of the SPES XT modeling framework, several study types were applied. For example, controlled experiments in academic settings were used to assess the impact and soundness of the approach, case studies were conducted on industrial example cases to evaluate applicability in industry, and industry surveys

were used to ensure appropriateness for industrial solution. Fig. 15-1 shows the structured combination of several empirical studies to evaluate the SPES XT modeling framework. Six exemplary industrial challenges, provided by industry based on their needs, were used to evaluate the SPES XT modeling framework's ability to cope with industry challenges. For each of these engineering challenges, academic evaluations as well as industrial studies from different engineering domains were conducted. In detail, evaluations such as industry surveys were conducted in the automotive, avionics, and automation domains. Case studies were applied to industrial case examples from these domains.

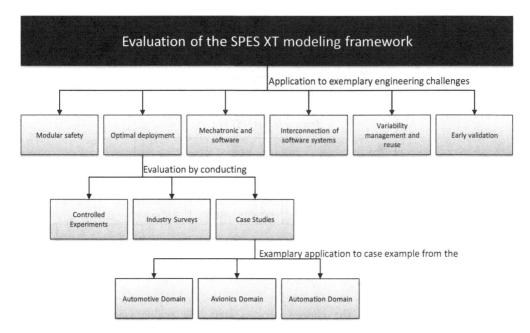

Fig. 15-1   *Combination of engineering challenges, study types, and engineering domains*

15.2   Evaluation Strategy

To ensure comparability of the evaluation results and allow single evaluation results to be aggregated into more general evidence, a step-by-step empirical research strategy was employed (cf. [Jedlitschka et al. 2013]), consisting of characterization, process choice, execution and analysis.

Characterize goals and strategies   In an initial *characterization* step, a hierarchy of all evaluation goals and related strategies was consolidated using GQM*Strategies® [Basili et al. 2014] (see Fig. 15-2). The high-level overall evaluation goals were then refined until atomic goals for single studies were derived. For more detailed examples, see [Lampasona et al. 2013].

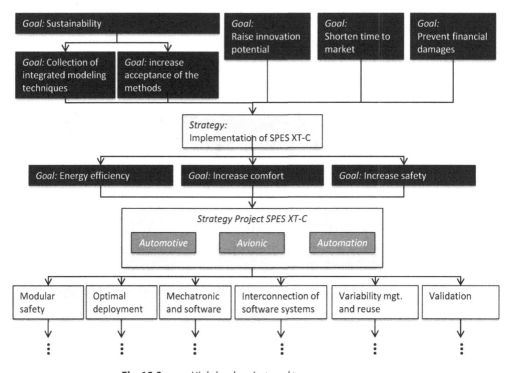

Fig. 15-2   *High-level project goal tree*

Choose evaluation method   Within the *choose process* step, the evaluations were defined in more detail, (i.e., the study type and data collection method(s) were chosen). In an industrial setting, case studies [Runeson et al. 2012] are often a preferred means using observations or surveys as data collection methods. In addition, the detailed design of each evaluation is defined by refining the very generic measurement goals into specific metrics used for the single evaluations with the help of GQM [Basili et al. 2001].

Execute study and analyze results   In the *execution and analysis* step, the design and summarized results of all the individual studies conducted for evaluating the SPES XT modeling framework were collected and aggregated to provide an overview of all evaluations. This collection includes all evaluation information, such as the study object, study type, measurement goal(s), and the case examples used.

15.3 Method Toolkit

A toolkit provided support for evaluations of the SPES XT modeling framework. This toolkit includes (1) descriptions of different evaluation types, (2) decision support for different study types, and (3) an overview of data collection methods to support the different study types.

In accordance with [Yin 2003], we can distinguish between different evaluation types: controlled experiment, case study, and industry survey. For each type of study, the toolkit provides a *detailed definition*, *usage scenario*, *degree of control*, *resources needed*, *data collection methods*, *data analysis methods*, *advantages,* and *disadvantages*. The decision support provided by [Schmidt 2006] is used to select a proper evaluation type based on *research questions* and the necessary *degree of control*.

Decision support

Based on the selected study type, up to six data collection methods can be used to collect evaluation data. Tab. 15-1 provides an overview of the usefulness of these collection methods for particular study types

Data collection method

Tab. 15-1   *Mapping data collection methods to study types*

Data collection method	Industry survey	Case study	Controlled experiment
Document analysis		++	++
Retrospective	+	++	
Single interview	++	++	++
Group interview	+	+	
Survey	+++	+	+++
Observation		+++	+++

15.4 Evaluation Landscape

While several studies with different study types have been conducted to ensure a holistic evaluation, a major concern of industry deals with the applicability of the SPES modeling framework to real industrial systems. As a consequence, much attention was given to the definition of real industrial case examples. These case examples were used in controlled experiments and in particular in case study research. Chapter 2 presented the running examples *Exterior Light-*

Case examples

ing and Speed Control (automotive) and *Desalination Plant* (automation), which were used for evaluation. In addition, several more case examples were used to ensure a broad evaluation.

Coverage of application areas

Tab. 15-2 gives a generic overview of the case study evaluations performed. The table indicates which case example has been applied to the different application challenges. This landscape shows that each example was used in the evaluation. The empirical evidence of the evaluations of the SPES XT modeling framework is underlined by the fact that each application area covers at least three different examples from different domains. Thus, the overall results for one application area are generalized. Tab. 15-2 also shows that the two running examples (see Chapter 2) are those used the most often and they allow seamless integration of the techniques developed, covering most of the development process.

Tab. 15-2   *Case example application in engineering challenges*

		Modular safety assessment	Optimal deployment	Mechatronic systems	Connected software systems	Variant management and reuse	Early validation
Automation	Desalination plant		X	X		X	X
Automotive	Electromagnetic braking system			X			
	Electronic engine control and emission handling			X		X	X
	Exterior lighting and speed control	X	X	X	X	X	X
	Lane-keeping assistance system	X	X		X		
Avionics	Cabin architecture optimization		X				
	Collision avoidance system	X			X		X
	Flight control system for civil aircraft		X				
	Flight control for unmanned aerial vehicles				X	X	

15.5 Applications of the Evaluation Strategy

In the following, a brief overview of the combination of exemplary studies conducted as part of the SPES XT modeling framework's overall evaluation is given.

The applicability of the SPES XT modeling framework in the industrial engineering challenge of modular safety assessment was evaluated in different studies. In particular, a multi-case study consisting of two case examples from the automotive (*Exterior Lighting and Speed Control*, Chapter 2) and avionics domains (*Collision Avoidance System*) was conducted in an industrial setting in the aforementioned domains. A further case study was conducted in the automation domain by applying the SPES XT modeling framework to the desalination plant [Brandstetter et al. 2015]. In summary, the SPES XT modeling framework can be applied in all three engineering domains to support modular safety assessment.

Modular safety assessment example

Another example for the combination of different studies is the use of empirical studies to evaluate the usefulness of the SPES XT modeling framework to deal with the challenges resulting from connected software systems. Several studies were conducted in this area. For example, an initial industry survey was conducted to evaluate the appropriateness of the methodological building blocks of the SPES XT modeling framework to support connected software systems development in the automotive and avionics domains [Daun et al. 2014a]. To evaluate the impact (by means of effectiveness, efficiency, user confidence, and supportiveness) of the proposed solution concepts, a controlled experiment, using different example cases as experiment material, was conducted [Daun et al. 2015]. Furthermore, applicability was evaluated by several case studies, for example, by the application of model-based documentation of behavioral requirements and functional design for the exterior lighting system [Föcker et al. 2015]. We can state that in combination, the proposed methodological building blocks of the SPES XT modeling framework can significantly increase the effectiveness and efficiency of the development and are also applicable and appropriate for industrial engineering processes.

Connected software systems example

In addition to the evaluations of the application challenges, crosscutting concerns were evaluated. For example, the suitability of the SPES XT modeling framework to support knowledge transfer was investigated. To do so, the SPES XT modeling framework's capability for technology transfer (i.e., the general application to

Crosscutting concerns example

industrial settings, cf. Chapter 13) and the ability to aid in teaching model-based engineering (cf. [Daun et al. 2014b]) were evaluated.

15.6 Summary

Because the individual evaluations predominantly showed positive results, we can infer that the different applications of the SPES XT modeling framework as well as its underlying modeling theory address the industrial challenges and provide a suitable solution for them. Considering the high-level goals in Fig. 15-2, the applicability of the SPES XT modeling framework has been confirmed in industrial settings.

Reliability of evaluation results   In particular, the reliability of the evaluation and the resulting evidence was ensured by the large number of different studies conducted (around 54 different studies). The evaluations showed that the acceptance and the usability of the proposed SPES XT modeling framework's techniques are higher when compared to previously used techniques.

Innovation potential   The innovation potential of the SPES XT modeling framework raised by the fruitful collaborations between different industry and academic partners address challenges posed by industry. The result is the previously mentioned large number of techniques. Some of the techniques developed, especially for modular safety assessment (Chapter 13) and early validation of engineering artifacts (Chapter 6), even helped to prevent financial damage.

15.7 References

[Basili 1992] V. Basili: The Experimental Paradigm in Software Engineering. Lecture on Experimental Software Engineering Issues: Critical Assessment and Future Directives, H.D. Rombach, V. Basili, and R. Selby, eds., Springer, 1993

[Basili et al. 2001] V. Basili, G. Caldiera, D. Rombach: Goal, question, metric paradigm. In: J. Marciniak (Eds.): Encyclopedia of Software Engineering, vol. 1, Wiley-Interscience, New York, pp. 511-519, 2001.

[Basili et al. 2014] V. Basili, A. Trendowicz, M. Kowalczyk, J. Heidrich, C. Seaman, J. Münch, D. Rombach: Aligning Organizations through Measurement – The GQM+Strategies Approach. Springer, Berlin Heidelberg, 2014.

[Brandstetter et al. 2015] V. Brandstetter, A. Froese, B. Tenbergen, A. Vogelsang, J.C. Wehrstedt, T. Weyer: Early Validation of Automation Plant Control Software using Simulation Based on Assumption Modeling and Validation Use Cases. Complex Systems Informatics and Modeling Quarterly, vol. 4, pp. 50-65, 2015.

[Daun et al. 2014a] M. Daun, J. Höfflinger, T. Weyer: Function-Centered Engineering of Embedded Systems - Evaluating Industry Needs and Possible Solutions.
9th Int. Conf. on Evaluation of Novel Approaches to SE, pp. 226-234, 2014.

[Daun et al. 2014b] M. Daun, A. Salmon, B. Tenbergen, T. Weyer, K. Pohl: Industrial case studies in graduate requirements engineering courses: The impact on student motivation. 27th IEEE Conf. on SE Education and Training, pp. 3-12,
2014.

[Daun et al. 2015] M. Daun, A. Salmon, T. Weyer, K. Pohl: The impact of students'
skills and experiences on empirical results: a controlled experiment with undergraduate and graduate students. 19th Int. Conf. on Evaluation and Assessment
in SE, paper 29, 2014.

[Föcker et al. 2015] F. Föcker, F. Houdek, M. Daun, T. Weyer: Model-Based Engi-
neering of an Automotive Adaptive Exterior Lighting System - Realistic Example
Specifications of Behavioral Requirements and Functional Design, ICB Research
Report, vol. 64, University of Duisburg-Essen, Essen, 2015.

[Jedlitschka et al. 2013] A. Jedlitschka, L. Guzman, J. Jung, C. Lampasona, S. Stein-
bach: Empirical Practice in Software Engineering. In: J. Münch, K. Schmid
(Eds.): Perspectives on the future of Software Engineering, Springer, Berlin Heidelberg, 2013.

[Lampasona et al. 2013] C. Lampasona, L. Guzman, P. Diebold: Document Analysis
as Extension of the GQM+Strategies® Approach to Support Evaluation. In: Proceedings of MetriKon 2013. Shaker Verlag, Aachen, 2013.

[Runeson et al. 2012] P. Runeson, M. Höst, A. Rainer, B. Regnell: Case Study Research in Software Engineering: Guidelines and Examples. Wiley, Hoboken, New
Jersey, 2012.

[Schmidt 2006] L. Schmidt: Technologie als Prozess. Eine empirische Untersuchung
organisatorischer Technologiegestaltung am Beispiel von Unternehmenssoftware.
PhD thesis. Berlin, 2006.

[Yin 2003] R. K. Yin: Case Study Research: Design and Methods. Thousand Oaks,
Vol. 5, No. 3, New York, 2003.

Manfred Broy

16

Outlook

In this chapter, we briefly summarize the project and its contributions as a whole. In addition, this chapter provides an outlook to open challenges in the engineering of software-intensive embedded systems and their impact on future research endeavors.

© Springer International Publishing AG 2016
K. Pohl et al. (eds.), *Advanced Model-Based Engineering of Embedded Systems*,
DOI 10.1007/978-3-319-48003-9_16

Seamless, model-based development

The area of model-based development is motivated and character-ized by a variety of pragmatic as well as scientifically well-founded methods. The SPES projects (i.e., SPES 2020 and SPES XT) have built a bridge between the more scientific approaches to modeling and the more pragmatic approaches, thus combining the benefits of both.

Using tools available on the market and providing the necessary solid scientific methodological support lays the foundation for engi-neering of embedded systems that can be extended in a variety of directions. In doing so, the SPES projects have integrated and con-solidated the different approaches from the application domains to form an integrated model-based approach that provides compre-hensive tool support and can be deployed in various application areas. SPES XT has conducted a number of very promising case studies in its application domains, thus showing how modeling can help in the development of powerful embedded systems. Moreover, we have demonstrated that the modeling techniques are useful not only for forward systems engineering but also for more specific tasks, such as functional safety or the verification of real-time be-havior. Finally, we successfully started to transfer these results into industrial practice. Therefore, at the end of the project, we can con-clude that the projects have reached their initial goal of implement-ing an integrated and seamless, model-based development approach for embedded systems.

As its main result, SPES XT contributes a broad and integrated concept which includes the following:

❏ Theoretical foundation
❏ Extensible framework and building blocks for system definition (SPES XT modeling framework and SPES XT Process Building Block Framework)
❏ Metrics for classification and assessment
❏ Tools and tool platform
❏ Guidelines and integrated process descriptions

Guiding principles

To achieve good integration of the results and optimal applicability in industrial practice, all solutions developed in SPES XT rely on four central principles:

❏ Principle 1: Comprehensiveness and ease of integration
❏ Principle 2: Development efficiency and verified quality
❏ Principle 3: Tool support
❏ Principle 4: Practical applicability

Because of the importance of these principles, SPES XT has introduced crosscutting subprojects in the project structure. These subprojects should guarantee the thorough and consistent implementation of these principles throughout the project by providing theoretical and technical frameworks as well as guidelines for their application. Most importantly, work in the crosscutting subprojects concentrated on the possibility to integrate the results — even in a changed context — in order to harmonize them and to objectively assess the quality of the artifacts created during the development process.

Based on this theoretical and methodological foundation, SPES XT has worked on solutions for a number of engineering challenges which are characterized by high practical relevance and complexity. The models of those engineering challenges have been seamlessly integrated into the basic framework. The practical applicability of the models and methods has been proven in numerous case studies.

Engineering challenges

Nevertheless, new trends, such as cyber-physical systems or autonomous systems, come with new challenges that are triggered by a number of key requirements of the field and go beyond the work and scope of the SPES projects.

Those systems are characterized by an open, networked character and can adapt flexibly to changing environments. We call them collaborating embedded systems. Collaborating embedded systems are part of a composite network of systems (composite system) which interact to provide functions that go beyond the functions of the individual members of the network. The new character of those systems calls for new architectures as well as a more specific methodology for the development of the systems. Examples for new challenges that arise from the development of collaborating embedded systems are:

Composite systems

❏ *Flexible architectures for collaborating embedded systems*: Dynamic environments (contexts), massive networking of collaborating systems, and the resulting demand for flexibility and adaptivity require new architecture designs and impact the development process of such systems. New architectures have to be flexible to deal with contexts that are not known at the time the architecture is designed, but rather have to be configurable at runtime. This includes self-adaptive, (semi-)autonomous system behavior as well as the possibility to change system components at runtime.

❏ *Open context*: This new challenge deals with the consequences of dynamic environments (i.e., contexts which are not precisely known at design time or may change during runtime) for the design of collaborating and composite systems. An individual system in this type of composite system typically only has a limited view of the real world as well as limited knowledge about the behavior and intentions of other members in the integrated network of systems. This type of fuzzy knowledge about the context also impacts system qualities such as safety, fault tolerance, or performance.

❑ *Human centric engineering*: The human factor plays a major role because the consequences of interactions between humans and systems are far-reaching. Therefore, the role of humans must be dealt with during the design phase of composite systems. There are two central problems that need to be addressed: (1) *Human-in-the-loop engineering at runtime* and (2) *Human-centric interface engineering for complex systems*.

❑ *Networking of collaborating embedded systems*: Collaborating embedded systems interact within networks of composite systems, which are dynamically built at system runtime, in order to provide additional value. Today, modeling and feature analysis of such systems is limited to the isolated view of individual systems in this type of network. Even if the individual embedded systems show the necessary system qualities, there is a risk that dynamic networking of those systems into integrated networks will not fulfill the intended purpose or user safety can no longer be guaranteed. The individual systems have to be developed in such a way that networking does not jeopardize the necessary system qualities.

Moving from classical embedded systems to networks of collaborating embedded systems opens up a huge new field of applications. At the same time, as discussed above, this transition comes with a quantum leap in complexity. Companies that are able to develop high-quality collaborating embedded systems will have a competitive advantage. To master this new complexity, practical methodological model-based approaches have to be developed. The results of the SPES projects form a perfect foundation towards this next step in complexity.

Appendices

A – Author Index

© Springer International Publishing AG 2016

K. Pohl et al. (eds.), *Advanced Model-Based Engineering of Embedded Systems*,

DOI 10.1007/978-3-319-48003-9

A

Albers, Dr. Karsten
INCHRON GmbH
Karl-Liebknecht-Str. 138   75,119,
14482 Potsdam, Germany   145,251

Antonino, Pablo
Chair of Software Engineering: Dependability
(SEDA)
Technische Universität Kaiserslautern
Postfach 3049
67653 Kaiserslautern, Germany   169

B

Bandyszak, Torsten
paluno – The Ruhr Institute for Software
Technology
University of Duisburg-Essen
Gerlingstr. 16
45127 Essen, Germany   241

Battram, Peter
BOOM SOFTWARE GMBH
Schneiderkruger Straße 12
49429 Visbek, Germany

Formerly:

Oldenburg Institute for Information Technology (OFFIS)
Escherweg 2
26121 Oldenburg, Germany   75,169

Beck, Stefan
Airbus Defence & Space
Rechliner Str.
85077 Manching, Germany   119

Bizik, Kai
Chair of Software Engineering: Dependa-
bility (SEDA)
Technische Universität Kaiserslautern
Postfach 3049
67653 Kaiserslautern, Germany   169

Bognar, Alfred
Airbus DS Electronic and Border Security
GmbH
Wörthstraße 85
89077 Ulm, Germany   75

Böhm, Birthe
Siemens AG
Corporate Technology
Research and Technology Center
Günther-Scharowsky-Str. 1
91058 Erlangen, Germany   225

Böhm, Dr. Wolfgang
Department of Informatics
Technische Universität München (TUM)
Boltzmannstr. 3
85748 Garching, Germany   37,59

Boyer, Arnaud
Airbus DS Electronic and Border Security
GmbH
Wörthstraße 85
89077 Ulm, Germany   169

Brandstetter, Veronika
Siemens AG
Otto-Hahn-Ring 6
81739 Munich, Germany   75

Brings, Jennifer
paluno – The Ruhr Institute for Software
Technology
University of Duisburg-Essen
Gerlingstr. 16
45127 Essen, Germany   43

Broy, Prof. Dr. Dr. h.c.
Manfred
Department of Informatics
Technische Universität München (TUM)
Boltzmannstr. 3
85748 Garching, Germany   iii,3,273

Büker, Dr. Matthias
Oldenburg Institute for Information Technology (OFFIS)
Escherweg 2
26121 Oldenburg, Germany   119,145

Buono, Suryo
Berner & Mattner Systemtechnik GmbH
Erwin-von-Kreibig-Str. 3
80807 Munich, Germany   169

C

Campetelli, Alarico
Department of Informatics
Technische Universität München (TUM)
Boltzmannstr. 3
85748 Garching, Germany   225

D

Daembkes, Prof. Dr. Heinrich
ARTEMIS Industry Association
High Tech Campus 69 - 3rd Floor
5656 AG Eindhoven, The Netherlands   iii,3

Daun, Marian
paluno – The Ruhr Institute for Software
Technology
University of Duisburg-Essen
Gerlingstr. 16   37,43,
45127 Essen, Germany   119,263

Davidich, Dr. Maria
Siemens AG
Otto-Hahn-Ring 6
81739 Munich, Germany   225

Diebold, Philipp
Fraunhofer Institute for Experimental Software Engineering (IESE)
Fraunhofer-Platz 1
67663 Kaiserslautern, Germany   241,263

Dieudonné, Laurent
Liebherr-Aerospace Lindenberg GmbH
Pfänderstraße 50-52
88161 Lindenberg, Germany   145

F

Froese, Dr. Andreas
paluno – The Ruhr Institute for Software
Technology
University of Duisburg-Essen
Gerlingstr. 16
45127 Essen, Germany   75

G

Gürsoy, Ömer
itemis AG
Am Brambusch 15
44536 Lünen, Germany   251

H

Henkler, Prof. Dr. Stefan
University of Applied Sciences Hamm-
Lippstadt
Marker Allee 76-78
59063 Hamm

Formerly:

Oldenburg Institute for Information Technology (OFFIS)
Escherweg 2
26121 Oldenburg, Germany   59,251

Heuer, André
paluno – The Ruhr Institute for Software
Technology
University of Duisburg-Essen
Gerlingstr. 16
45127 Essen, Germany   197,241,263

Hilbrich, Dr. Robert
Deutsches Zentrum für Luft- und Raum-
fahrt (DLR)
Rutherfordstraße 2
12489 Berlin

Formerly:

Das Fraunhofer-Institut für Offene Kom-
munikationssysteme FOKUS
Kaiserin-Augusta-Allee 31
10589 Berlin   145

Himsolt, Dr. Michael
Daimler AG
Research & Development
Wilhelm-Runge-Str. 11
89081 Ulm, Germany   197

Hönninger, Harald
Robert Bosch GmbH
Corporate Research
P.O. Box 30 02 40
70442 Stuttgart, Germany   iii,3

Houdek, Dr. Frank
Daimler AG
Research & Development
Wilhelm-Runge-Str. 11
89081 Ulm, Germany   11,59

282   Authors

I

Igna, Dr. Georgeta
fortiss GmbH
Guerickestr. 25
80805 Munich, Germany   145

J

Jäger, Dr. Tobias
Siemens AG
Corporate Technology
Research and Technology Center
Frauenauracher Str. 80
91056 Erlangen, Germany   197

Jedlitschka, Dr. Andreas
Fraunhofer Institute for Experimental
Software Engineering (IESE)
Fraunhofer-Platz 1
67663 Kaiserslautern, Germany   263

K

Kaiser, Dr. Bernhard
Berner & Mattner Systemtechnik GmbH
Erwin-von-Kreibig-Str. 3
80807 Munich, Germany   169

Käßmeyer, Michael
Audi Electronics Venture GmbH
Sachsstr. 20
85080 Gaimersheim, Germany   169

Kaufmann, Tobias
paluno – The Ruhr Institute for Software
Technology
University of Duisburg-Essen
Gerlingstr. 16
45127 Essen, Germany   197

Koutsoumpas, Vasileios
Department of Informatics
Technische Universität München (TUM)
Boltzmannstr. 3
85748 Garching, Germany   37

Kugele, Dr. Stefan
Department of Informatics
Technische Universität München (TUM)
Boltzmannstr. 3
85748 Garching, Germany   145

Kuhn, Dr. Thomas
Fraunhofer Institute for Experimental
Software Engineering (IESE)
Frauenhofer-Platz 1
67663 Kaiserslautern, Germany   145,241

L

Löwen, Dr. Ulrich
Siemens AG
Corporate Technology
Research and Technology Center
Günther-Scharowsky-Str. 1
91058 Erlangen, Germany   11,225

M

MacGregor, John
Robert Bosch GmbH
Siemensstr.
71701 Schwieberdingen, Germany   119

Manz, Christian
Institute of Databases and Information
Systems,
University Ulm
James-Franck-Ring 1
89081 Ulm, Germany   197

May, Reinhold
Airbus Defense and Space
Woerthstr. 85
89077 Ulm, Germany   197

O

Oehlerking, Dr. Jens
Robert Bosch GmbH
Robert-Bosch-Campus 1
71272 Renningen, Germany   105

P

Pohl, Prof. Dr. Klaus
paluno – The Ruhr Institute for Software
Technology
University of Duisburg-Essen
Gerlingstr. 16
45127 Essen, Germany   iii,3

Prohaska, Alexander
Chair of Software Engineering: Dependa-
bility (SEDA)
Technische Universität Kaiserslautern
Postfach 3049
67653 Kaiserslautern, Germany   169

R

Reuter, Christian
Daimler AG
Research & Development
Benz-Str.
71063 Sindelfingen, Germany   197

Rosinger, Maike
Oldenburg Institute for Information Technology (OFFIS)
Escherweg 2
26121 Oldenburg, Germany   145

Rumpe, Prof. Dr. Bernhard
Software Engineering Department of
Computer Science 3
RWTH Aachen University
Ahornstr. 55
52074 Aachen, Germany   197

S

Salmon, Andrea
paluno – The Ruhr Institute for Software
Technology
University of Duisburg-Essen
Gerlingstr. 16
45127 Essen, Germany   119

Schaefer, Prof. Dr. Ina
Institiute of Software Engineering and Auto-
motive Informatics
TU Braunschweig
Mühlenpfordtstr. 23
38106 Braunschweig   197

Schlingloff, Prof. Dr. Holger
Fraunhofer Institute for Computer Architecture
and Software Technology (FIRST)
Kekuléstr. 7
12489 Berlin, Germany   197

Schröck, Sebastian
Automation Technology Institute
Helmut Schmidt University Hamburg
Holstenhofweg 85
22043 Hamburg, Germany   197

Schulze, Christoph
Software Engineering Department of Com-
puter Science 3
RWTH Aachen University
Ahornstr. 55
52074 Aachen, Germany   197

Schulze, Dr. Michael
Pure-systems GmbH
Otto-von-Guericke-Str.28
39104 Magdeburg, Germany   197,251

Strathmann, Thomas
Oldenburg Institute for Information Technology (OFFIS)
Escherweg 2
26121 Oldenburg, Germany   105

T

Tenbergen, Bastian
Department of Computer Science
State University of New York at Oswego
Oswego, NY, 13126, USA

Formerly:

paluno – The Ruhr Institute for Software
Technology
University of Duisburg-Essen
Gerlingstr. 16   43,75,
45127 Essen, Germany   169,251

Terfloth, Axel
itemis AG
Am Brambusch 15
44536 Lünen, Germany   251

V

Velasco, Santiago
Fraunhofer Institute for Experimental
Software Engineering (IESE)
Frauenhofer-Platz 1
67663 Kaiserslautern, Germany   169

Vetrò, Dr. Antonio
Department of Informatics
Technische Universität München (TUM)
Boltzmannstr. 3
85748 Garching, Germany   241

Vogelsang, Dr. Andreas
Department of Informatics
Technische Universität München (TUM)
Boltzmannstr. 3
85748 Garching, Germany   37,59,75

W

Weber, Raphael
Timing-Architects Embedded Systems
GmbH
Franz-Mayer-Strasse 1
93053 Regensburg

Formerly:

Oldenburg Institute for Information Technology (OFFIS)
Escherweg 2
26121 Oldenburg, Germany   119,145,251

Wegener, Dr. Joachim
Berner & Mattner Systemtechnik GmbH
Erwin-von-Kreibig-Str. 3
80807 Munich, Germany   75

Wehrstedt, Dr. Jan Christoph
Siemens AG
Otto-Hahn-Ring 6
81739 Munich, Germany   11

Weißleder, Dr. Stephan
Das Fraunhofer-Institut für Offene Kom-
munikationssysteme FOKUS
Kaiserin-Augusta-Allee 31
10589 Berlin   197

Weyer, Dr. Thorsten
paluno – The Ruhr Institute for Software
Technology
University of Duisburg-Essen
Gerlingstr. 16   37,43,59,
45127 Essen, Germany   119,241

Z

Zander, Dr. Justyna
Berner & Mattner Systemtechnik GmbH
Erwin-von-Kreibig-Str. 3
80807 Munich, Germany   169

Zimmer, Florian
Helmut-Schmidt-University
University of Bundeswehr Hamburg
Holstenhofweg 85
22043 Hamburg, Germany   225

B – Project Structure

© Springer International Publishing AG 2016
K. Pohl et al. (eds.), *Advanced Model-Based Engineering of Embedded Systems*,
DOI 10.1007/978-3-319-48003-9

Co-operation

The project focused on a strong interleaving of science and praxis, in order to make good progress in the engineering challenges, to verify the approaches and to transfer the results in engineering methods that are viable in the praxis. This target has been supported by leveraging the existing partner network from the predecessor project SPES 2020 which has been systematically expanded by new partners.

The structure between the different locations in Germany allowed for the inclusion of key resources into the project. The close co-operation between the industrial partners and the universities/Fraunhofer institutes has made sure that the results could be evaluated with respect to practical and market feasible solutions.

Based on this partner network we have designed the SPES XT project structure with specific focus to foster tight integration of results throughout the project between engineering challenges as well as across application domains.

Project Structure

Leveraging on the rich results of the predecessor project SPES 2020, SPES XT has consistently extended the SPES 2020 methodology. Because of the increasing importance of various engineering problems, the industry partners have defined six advanced challenges in the development of embedded systems which have been addressed within SPES XT.

The SPES XT project structure logically structures the rich problem and solution space of the project. As shown in Fig. C we have implemented a 3-dimensional project structure with the engineering challenges (EC) as the central element, which were addressed in cross domain project activities. Orthogonal work on cross-cutting topics (QT) was supposed to support integrated and seamless results which can easily be applied in industrial practice. In the third dimension from the application domains, domain specific characteristics were addressed.

Dimension "Engineering Challenges"

Six engineering challenges have been defined by the industry partners which structured the work from a technical engineering point of view.

- ❑ EC1: Modular Safety Case
- ❑ EC2: Optimal Deployment
- ❑ EC3: Mechatronic and Software
- ❑ EC4: Networks
- ❑ EC5: Variant Management and Reuse

❑   EC6: Validation in early Phases

These engineering challenges are mutually not independent. This leads to desired cross references and cooperation between the sub-projects which in turn had positive effects on the overall methodical concept.

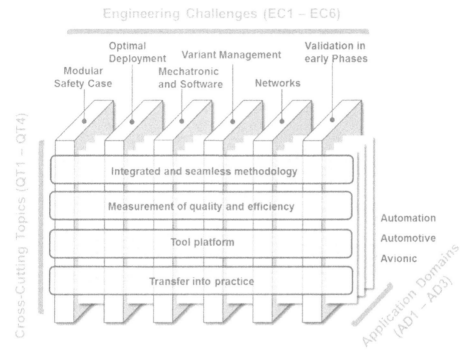

Fig. C   *SPES XT project structure*

Dimension "Cross-Cutting Topics"

In order to support comprehensiveness of the results and to guarantee practical applicability, all SPES XT results are based on four central principles:

❑   Principle 1: Integrated and Seamless Methodology
❑   Principle 2: Measurement of Quality and Efficiency
❑   Principle 3: Tool Support
❑   Principle 4: Transfer into Practice

Due to the importance of these principles we established cross-cutting topics in the project structure orthogonal to the engineering challenges to guarantee systematic and consistent consideration.

In the cross-cutting projects work was done to harmonize the results and to make sure that they can be seamlessly integrated even in a different context.

Dimension "Application Domains"

We applied the results to three application domains (AD) which built the third dimension of the project structure:

❑ Automation
❑ Automotive
❑ Avionics

Thereby we extended the scope of the engineering challenges and address necessary domain specific characteristics (e.g., safety, formal certification, real time requirements, etc.). Because of these characteristics the engineering challenges may have different manifestations which in turn led to variants in the developed solutions.

C – Members of the SPES XT Project

© Springer International Publishing AG 2016
K. Pohl et al. (eds.), *Advanced Model-Based Engineering of Embedded Systems*,
DOI 10.1007/978-3-319-48003-9

Members	Website
Airbus Operations GmbH	http://www.airbus.com/
Audi Electronics Venture	http://www.audi-electronics-venture.de/
Berner & Mattner GmbH	http://www.berner-mattner.com/
Airbus Defence & Space GmbH	http://airbusdefenceandspace.com/
Daimler AG	http://www.daimler.com/
EADS-Deutschland GmbH	http://www.airbusgroup.com/
fortiss GmbH	http://www.fortiss.org/
Fraunhofer-Institut für Experimentelles Software Engineering	http://www.iese.fraunhofer.de/
Fraunhofer-Institut für offene Kommunikations-systeme FOKUS	https://www.fokus.fraunhofer.de/
Helmut-Schmid-Universität	http://www.hsu-hh.de/
Inchron GmbH	https://www.inchron.de/
Itemis AG	http://www.itemis.de/
Liebherr Aerospace Lindenberg GmbH	http://www.liebherr.com/
OFFIS e.V.	http://www.offis.de/
Pure-Systems GmbH	http://www.pure-systems.com/
Rheinisch-Westfälische Technische Hochschule Aachen	http://www.rwth-aachen.de/
Robert Bosch GmbH	http://www.bosch.de/
Siemens AG	http://www.siemens.com/
University of Kaiserslautern	https://www.uni-kl.de/
Technische Universität München (TUM)	https://www.tum.de/
University of Duisburg-Essen	https://www.uni-due.de/

D – List of Publications

© Springer International Publishing AG 2016
K. Pohl et al. (eds.), *Advanced Model-Based Engineering of Embedded Systems*,
DOI 10.1007/978-3-319-48003-9

A

[Aßmann et al. 2014] U. Aßmann, S. Götz, J-M. Jézéquel, B. Morin, M. Trapp: A Reference Architecture and Roadmap for Models@run.time Systems. In: Models@run.time. Foundations, Applications, and Roadmaps, Springer, 2014, pp. 1-18.

B

[Bender et al. 2015] O. Bender, W. Böhm, F. Houdek, S. Henkler, A. Vogelsang, T. Weyer: Fünfter Workshop zur Zukunft der Entwicklung eingebetteter Systeme (ENVISION 2020). In: Proceedings of Software Engineering & Management 2015, Lecture Notes in Informatics, GI Edition, 2015.

[Böhm and Vogelsang 2013] W. Böhm, A. Vogelsang: An Artifact-oriented Framework for the Seamless Development of Embedded Systems. In: Workshopband Software Engineering 2013, Lecture Notes in Informatics (LNI), GI-Edition, Dritter Workshop zur Zukunft der Entwicklung softwareintensiver eingebetteter Systeme (ENVISION2020), 2013, pp. 225 - 234.

[Böhm and Junker 2015] W. Böhm, M. Junker: Siemens Rail - Industrial Case Study: Model-based Development of a Train Guard MT Function. In: Proceedings of Software Engineering & Management 2015, Lecture Notes in Informatics, GI Edition, 2015.

[Böhm et al. 2014] W. Böhm, S. Henkler, F. Houdek, A. Vogelsang, T. Weyer: Bridging the Gap between Systems and Software Engineering by Using the SPES Modeling Framework as a General Systems Engineering Philosophy. In: Elsevier Procedia Computer Science, Vol. 28, 2014, pp. 187 - 194.

[Brandstetter et al. 2013] V. Brandstetter, J. C. Wehrstedt, R. Rosen, A. Pirsing: Simulationsgestütze Entwicklung von Automatisierungssoftware. In: ATP-Edition, Fachmagazin für Automatisierungstechnische Praxis 06/2013, 2013.

[Brandstetter et al. 2013] V. Brandstetter, J. C. Wehrstedt, R. Rosen, A. Pirsing: Simulationsgestütze Entwicklung von Automatisierungssoftware. In: ATP-Edition, Fachmagazin für Automatisierungstechnische Praxis 06/2013, 2013.

[Brandstetter et al. 2015] V. Brandstetter, A. Froese, B. Tenbergen, A. Vogelsang, J. C. Wehrstedt, T. Weyer: Early Validation of Automation Plant Control Software using Simulation Based on Assumption Modeling and Validation Use Cases. Complex Systems Informatics and Modeling Quarterly (CSIMQ), Vol. 4, 2015.

[Brandstetter et al. 2015] V. Brandstetter, A. Froese, B. Tenbergen, A. Vogelsang, J. C. Wehrstedt, T. Weyer: Early Validation of Control Software for Automation Plants on the Example of a Seawater Desalination Plant. 27th International Conference on Advanced Information Systems Engineering (CAiSE), Forum, Stockholm, Schweden, 2015.

[Broy 2013] M. Broy: Model-Based Software & Systems Engineering – Abstraction and Structuring. Keynote at ENVISION 2013, Aachen, 2013.

[Bücker et al. 2013] M. Bücker, S. Henkler, S. Schlegel, E. Thaden: A Design Space Exploration Framework for Model-Based Software-intensive Embedded System Development. In: Workshopband Software Engineering 2013, GI-Edition Lecture Notes in Informatics (LNI), Dritter Workshop zur Zukunft der Entwicklung softwareintensiver eingebetteter Systeme (ENVISION2020), 2013, pp. 245 - 250.

[Bücker et al. 2013] M. Büker, W. Damm, G. Ehmen, S. Henkler, D. Janssen, I. Stierand, E. Thaden: From Specification Models to Distributed Embedded Applications: A Holistic User-Guided Approach. SAE International Journal of Passenger Cars- Electronic and Electrical Systems, 2013.

[Bücker et al. 2013] M. Büker, S. Henkler, S. Schlegel, E. Thaden: A Design Space Exploration Framework for Model-Based Software-intensive Embedded System Development. 2013.

[Bücker et al. 2013] M. Büker, W. Damm, G. Ehmen, S. Henkler, A. Rettberg, I. Stierand, E. Thaden: A
   Design Space Exploration Tool Demonstration for Automotive Systems. DAC, 2013.

C

[Campitelli 2013] A. Campitelli: Dynamic Sampling for Focus Hybrid Components. In: International
   Journal of Engineering and Technology (IJET), 2nd International Conference on Civil Engineering
   and Architecture, Barcelona, 2013.

[Campitelli et al. 2015] A. Campetelli, M. Junker, V. Koutsoumpas, X. Zhu, B. Böhm, M. Davidich J. C.
   Wehrstedt: Realizing Complex Model Transformations with QVTo: Two Case Examples. In: Fünfter
   Workshop zur Zukunft der Entwicklung eingebetteter Systeme (ENVISION 2020), 2015.

[Campitelli and Hackenberg 2015] A. Campetelli, G. Hackenberg: Performance Analysis of Adaptive
   Runge-Kutta Methods in Region of Interest. In: 2nd International IFIP Workshop on Emerging Ideas
   and Trends in Engineering of Cyber-Physical Systems (EITEC '15), 2015.

[Cerone et al. 2013] A. Cerone, M. Roggenbach, H. Schlingloff, G. Schneider, S. Shaikh: Teaching Formal Methods for Software Engineering - Ten Principles. FWFM 2013 - Fun With Formal Methods
   Workshop on 25th International Conference on Computer Aided Verification (CAV 2013), Russia,
   2013.

D

[Daun et al. 2013] M. Daun, J. Brings, J. Höfflinger, T. Weyer: Funktionsgetriebene Entwicklung soft-
   ware-intensiver eingebetteter Systeme in der Automobilindustrie – Stand der Wissenschaft und Forschungsfragestellungen. In: Workshopband Software Engineering 2013, GI-Edition Lecture Notes in
   Informatics (LNI), Dritter Workshop zur Zukunft der Entwicklung softwareintensiver eingebetteter
   Systeme (ENVISION2020), 2013, pp. 293 – 302.

[Daun et al. 2014] M. Daun, J. Brings, B. Tenbergen, T. Weyer: On the Model-based Documentation of
   Knowledge Sources in the Engineering of Embedded Systems. Vierter Workshop zur Zukunft der
   Entwicklung softwareintensiver eingebetteter Systeme (ENVISION2020), 2014.

[Daun et al. 2014] M. Daun, J. Höfflinger, T. Weyer: Function-Centered Engineering of Embedded Systems – Evaluating Industry Needs and Possible Solutions.In: Proceedings of 9th International Conference on Evaluation of Novel Approaches to Software Engineering (ENASE 2014), 2014.

[Daun et al. 2014] M. Daun, A. Salmon, B. Tenbergen, T. Weyer, K. Pohl: Industrial Case Studies in
   Graduate Requirements Engineering Courses: The Impact on Student Motivation. In: Proceedings of
   27th International Conference on Software Engineering Education and Training (CSEE&T 2014),
   2014.

[Daun et al. 2014] M. Daun, T. Weyer, K. Pohl: Validating the Functional Design of Embedded Systems
   Against Stakeholder Intentions. In: Proceedings of the 2nd International Conference on Model-
   Driven Engineering and Software Development, 2014.

[Daun et al. 2015] M. Daun, J. Brings, T. Bandyszak, P. Bohn, T. Weyer: Collaborating Multiple System
   Instances of Smart Cyber-Physical Systems: A Problem Situation, Solution Idea, and Remaining Research Challenges. International Workshop on Software Engineering for Smart Cyber-Physical Systems (SEsCPS), International Conference on Software Engineering (ICSE 2015), Florence, Italien.
   ACM, New York, 2015.

[Daun et al. 2015] M. Daun, A. Salmon, T. Weyer, K. Pohl: The Impact of Students' Skills and Experiences on Empirical Results: A Controlled Experiment with Undergraduate and Graduate Students. In:
   Proceedings of the19th International Conference on Evaluation and Assessment in Software Engineering (EASE), Nanjing, China. ACM, New York, 2015.

[Daun et al. 2015] M. Daun, A. Salmon, T. Weyer: Using dedicated Review Diagrams to detect Defective Functional Interplay in Function-Centered Engineering. In: Gemeinsamer Tagungsband der Workshops der Tagung Software Engineering, Fünfter Workshop zur Zukunft der Entwicklung softwareintensiver, eingebetteter Systeme (ENVISION 2020), CEUR Workshop Proceedings, 1337, 2015.

[Daun et al. 2015] M. Daun, B. Tenbergen, J. Brings, T. Weyer: Documenting Assumptions about the Operational Context of Long-Living Collaborative Embedded Systems. In: Gemeinsamer Tagungsband der Workshops der Tagung Software Engineering 2015, 2nd Workshop on Evolution and Maintenance of Long-Living Software Systems (EMLS 2015), CEUR Workshop Proceedings, 1337, 2015.

[Daun et al. 2015] M. Daun, T. Weyer, K. Pohl: Detecting and Correcting Outdated Requirements in Function-Centered Engineering of Embedded Systems. In: Proceedings of the 21th International Working Conference on Requirements Engineering – Foundation for Software Quality (REFSQ 2015), Lecture Notes in Computer Science, 9013, Springer, 2015, pp. 65-80.

[Daun et al. 2016] M. Daun, J. Brings, T. Weyer, B. Tenbergen: Fostering Concurrent Engineering of Cyber-physical Systems – A Proposal for an Ontological Context Framework: In: 3rd International Workshop on Emerging Ideas and Trends in Engineering of Cyber-Physical Systems (EITEC), IEEE Computer Society, Los Alamitos, 2016, pp. 5-10.

[Daun et al. 2016] M. Daun, P. Bohn, J. Brings, T. Weyer: Structured Model-Based Engineering of Long-living Embedded Systems: The SPES Methodological Building Blocks Framework. Softwaretechnik-Trends, Vol. 36, No. 1, 2016.

[Diebold and Vetrò 2014] P. Diebold, A. Vetrò: Bridging the gap: SE technology transfer into practice: study design and preliminary results. In: Proceedings of the 8th ACM/IEEE International Symposium on Empirical Software Engineering and Measurement (ESEM '14). ACM, New York, NY, USA, Article 52, 2014.

[Diebold and Vetrò 2015] P. Diebold, A. Vetrò: Characterizing Technology Transfer in Software Engineering: An Exploratory Survey. International Conference on Evaluation and Assessment in Software Engineering (EASE), 2015.

[Diebold et al 2015] P. Diebold, A. Vetrò, D. M. Fernandez: An Exploratory Study on Technology Transfer in Software Engineering. In: Empirical Software Engineering and Measurement (ESEM), 2015.

E

[Ehmen et al. 2013] G. Ehmen, A. Rettberg M. Büker, S. Henkler, E. Thaden: From MATLAB-SIMULINK To Distributed Embedded Applications: An Automotive Tool Demonstration. DATE, 2013.

F

[Filippidis and Schlingloff 2012] L. Filippidis, H. Schlingloff: Structural Equation Modelling for Causal Analysis Applied to Transport Systems. In: FORMS/FORMAT 2012, 9th International Symposium on Formal Methods, Braunschweig, 2012.

[Föcker et al. 2015] F. Föcker, F. Houdek, M. Daun, T. Weyer: Model-based Engineering of an Automotive Adaptive Exterior Lighting System – Specification of Behavioral Requirements and Functional Design. ICB-Research Report, No. 62, Essen, 2015.

G

[Godesa and Hilbrich 2013] J. Godesa, R. Hilbrich: Framework für die empirische Bestimmung der Ausführungszeit auf Mehrkernprozessoren. In: Proceedings des Workshop Echtzeit 2013 (Boppard, Deutschland), 21/22.11.2013, 2013.

[Godesa 2013] J. Godesa: Framework für die empirische Bestimmung der Ausführungszeit auf Mehr-kernprozessoren. Diploma Thesis, Humboldt-Universität Berlin, 2013.

[Göring et al. 2012] M. Göring, L. Christiansen, T. Holm, T. Jäger, A. Fay: ISO 15926 vs. IEC 62424 - Gegenüberstellung zweier Standards zur Modellierung von Anlagenstrukturen. In: Jahrestreffen der Fachgemeinschaft Prozess-, Apparate- und Anlagentechnik, Dortmund, 2012.

[Grigoleit 2015] F. Grigoleit, A. Vetrò, P. Diebold, D. M. Fernández, W. Boehm: In Quest for proper Mediums for Technology Transfer in Software Engineering – An Exploratory Study. In: Empirical Software Engineering and Measurement (ESEM), 2015.

[Große-Rhode et al. 2013] M. Große-Rhode, R. Hilbrich, S. Weißleder: Recommendation Systems for the Construction and Configuration of Product Lines. In: Recommendation Systems for Software Engineering, 2013.

[Große-Rhode et al. 2013] M. Große-Rhode, P. Manhart, R. Mauersberger, S. Schröck, M. Schulze, T. Weyer: Anforderungen von Leitbranchen der deutschen Industrie an Variantenmanagement und Wiederverwendung und daraus resultierende Forschungsfragestellungen. In: Workshopband Software Engineering 2013, GI-Edition Lecture Notes in Informatics (LNI), Dritter Workshop zur Zukunft der Entwicklung softwareintensiver eingebetteter Systeme (ENVISION2020), 2013, pp. 251 – 260.

[Große-Rhode et al. 2013] M. Große-Rhode, R. Hilbrich, S. Weißleder: Achieving System Quality in Safety-Critical Embedded Systems by Combining Automated Design Space Exploration, Validation, and Product Line Engineering. In: System Quality and Software Architecture (SQSA), 2013.

H

[Heuer et al. 2013] A. Heuer, T. Kaufmann, T. Weyer: Extending an IEEE 42010-compliant Viewpoint-based Engineering-Framework for Embedded Systems to Support Variant Management. In: Embedded Systems: Design, Analysis and Verification, IFIP Advances in Information and Communication Technology, Vol. 403, Springer, Heidelberg 2013, pp. 283 – 292.

[Heuer et al. 2013] A. Heuer, P. Diebold, T. Bandyszak: Supporting Technology Transfer by Providing Recommendations for Writing Structured Guidelines. Vierter Workshop zur Zukunft der Entwicklung softwareintensiver eingebetteter Systeme (ENVISION2020), 2014.

[Heuer and Pohl 2013] A. Heuer, K. Pohl: Structuring Variability in the Context of Embedded Systems during Software Engineering. In: Proceedings of the 8th International Workshop on Variability Modelling of Software-intensive Systems (VAMOS) 2014, ACM Press, 2014.

[Hilbrich 2012] R. Hilbrich: How to Safely Integrate Multiple Applications on Embedded Many-Core Systems by Applying the "Correctness by Construction" Principle. Advances in Software Engineering, Vol. 2012, 2012.

[Hilbrich and Dieudonné 2013] R. Hilbrich, L. Dieudonné: Deploying Safety-Critical Applications on Complex Avionics Hardware Architectures. In: Journal of Software Engineering and Applications, Vol. 6, No. 5, 2013, pp. 229 – 235.

[Holdschick 2014] H. Holdschick: Konzepte zur Absicherung der Variantenkonfiguration von eingebetteter Fahrzeugsoftware. Vierter Workshop zur Zukunft der Entwicklung softwareintensiver eingebetteter Systeme (ENVISION2020) 2014.

[Holm et al. 2012] T. Holm, L. Christiansen, M. Göring, T. Jäger, A. Fay: ISO 15926 vs. IEC 62424 - Comparison of Plant Structure Modeling Concepts. In: Proceedings of 17th International Conference on Emerging Technologies and Factory Automation (ETFA), Kraków, Poland, 2012.

[Holm 2013] T. Holm, S. Schröck, A. Fay, T. Jäger, U. Löwen: Engineering von "Mechatronik und Software" in automatisierten Anlagen: Anforderungen und Stand der Technik. In: Workshopband Software Engineering 2013, GI-Edition Lecture Notes in Informatics (LNI), Dritter Workshop zur

Zukunft der Entwicklung softwareintensiver eingebetteter Systeme (ENVISION2020), 2013, pp. 261 – 272.

J

[Jäger 2014] T. Jäger: Benefits of Process Modeling within the Engineering of Automated Facilities. In: ICIT 2014 – International Conference on Industrial Technology, Busan, South Korea, 2014.

[Junker 2014] M. Junker: Exploiting Behavior Models for Availability Analysis of Interactive Systems. International Symposium Software Reliability Engineering Workshops (ISSREW), 2014.

K

[Käßmeyer et al. 2015] M. Käßmeyer, D. S. V. Moncada, M. Schurius: Evaluation of a Systematic Approach in Variant Management for Safety-Critical Systems Development. 13th International Conference on Embedded and Ubiquitous Computing (EUC), 2015.

[Käßmeyer et al. 2015] M. Käßmeyer, M. Schulze, M. Schurius: A Process to Support a Systematic Change Impact Analysis of Variability and Safety in Automotive Functions. SPLC 2015, Nashville USA, 2015.

[Kaufmann et al. 2014] T. Kaufmann, C. Manz, T. Weyer: Extending the SPES Modeling Framework for Supporting Role-specific Variant Management in the Engineering Process of Embedded Software. Vierter Workshop zur Zukunft der Entwicklung softwareintensiver eingebetteter Systeme (ENVISION2020), 2014.

[Knapp et al. 2013] A. Knapp, L. O'Reilly, M. Roggenbach, H. Schlingloff: Simulating Timed UML2 Sequence Diagrams with Timed CSP. In: Electronic Communications of the EASST, Vol. X (2013). AVoCS 2013, Guildford, Surrey, UK, 2013.

[Kugele and Pucea 2014] S. Kugele, G. Pucea: Model-based Optimization of Automotive E/E-Architectures. In: Proceedings of 6th International Workshop on Constraints in Software Testing, Verification, and Analysis, CSTVA 2014, ACM, Hyderabad, India, 2014, pp. 18 – 29.

[Kugele et al. 2015] S. Kugele, G. Pucea, R. Popa, L. Dieudonné, H. Eckardt: On the Deployment Problem of Embedded Systems. In: Proceedings of the 13th ACM-IEEE International Conference on Formal Methods and Models for System Design (MEMOCODE'15), IEEE, Austin, USA, 2015, pp. 158 – 167.

L

[Liggesmeyer and Trapp 2014] P. Liggesmeyer, M. Trapp: Safety: Herausforderungen und Lösungsansätze. In: Industrie 4.0, Springer, 2014.

M

[Manhart et al. 2013] P. Manhart, P. M. S. Nazari, B. Rumpe, I. Schaefer, C. Schulze: Konzepte zur Erweiterung des SPES Meta-Modells um Aspekte der Variabilitäts- und Deltamodellierung. In: Workshopband Software Engineering 2013, GI-Edition, Lecture Notes in Informatics (LNI), Dritter Workshop zur Zukunft der Entwicklung softwareintensiver eingebetteter Systeme (ENVISION2020), 213, pp. 283 – 292.

[Manz and Reichert 2013] C. Manz, M. Reichert: Herausforderungen an ein durchgängiges Variantenmanagement in Software-Produktlinien und die daraus resultierende Entwicklungsprozessadaption. In: Workshopband Software Engineering 2013, GI-Edition, Lecture Notes in Informatics (LNI), Dritter Workshop zur Zukunft der Entwicklung softwareintensiver eingebetteter Systeme (ENVISION2020), 2013, pp. 273 – 282.

[Manz et al. 2014] C. Manz, M. Schulze, M. Reichert: An Approach to Detect the Origin and Distribution of Software Defects in an Evolving Cyber-Physical System. In: Workshop on Emerging Ideas and Trends in Engineering of Cyber-Physical Systems (EITEC '14), Berlin, Germany, 2014.

[Ming and Schlingloff 2013] C. Ming, H. Schlingloff: A Rewriting Based Monitoring Algorithm for TPTL. In: Proceedings of 22nd CS&P 2013 – Concurrency, Specification and Programming. Warsow, 2013, pp. 61 – 72.

[Ming and Schlingloff 2014] C. Ming, H. Schlingloff: Online Monitoring of Distributed Systems with a Five Valued LTL. In: Proceedings of ISVML 2014 – IEEE International Symposium on Multiple-Valued Logic. Bremen, 2014.

O

[Oertel et al. 2014] M. Oertel, A. Mahdi, E. Böde, A. Rettberg: Contract-based Safety: Specification and Application Guidelines. In: Workshop on Emerging Ideas and Trends in Engineering of Cyber-Physical Systems (EITEC '14), Berlin, Germany, 2014.

P

[Papendieck and Schulze 2014] M. Papendieck, M. Schulze: Concepts for Consistent Variant-Management Tool Integrations. Vierter Workshop zur Zukunft der Entwicklung softwareintensiver eingebetteter Systeme (ENVISION2020), 2014.

R

[Ramich 2014] V. Ramich: Teilautomatische Erstellung von Component-Fault-Trees aus Simulink-Modellen. Master Thesis, Universität Kassel, 2014.

[Kampenhout and Hilbrich 2013] J. R. van Kampenhout, R. Hilbrich: Model-Based Deployment of Mission-Critical Spacecraft Applications on Multicore Processors. In: Proceedings of Ada-Europe International Conference on Reliable Software Technologies 2013 (Berlin), LNCS 7896, Springer-Verlag Berlin Heidelberg, 2013, pp. 35 – 50.

[Reinkemeier et al. 2014] P. Reinkemeier, H. Hille, S. Henkler: Towards Creating Flexible Tool Chains for the Design and Analysis of Multi-Core Systems. Vierter Workshop zur Zukunft der Entwicklung softwareintensiver eingebetteter Systeme (ENVISION2020), 2014.

[Rosinger et al. 2014] M. Rosinger, M. Büker, R. Weber: A User-Supported Approach to Determine the Importance of Optimization Criteria for Design Space Exploration. CPSWeek 2014 – IDEAL'14 Workshop, 2014.

S

[Schlingloff 2014] H. Schlingloff: Towards a Curriculum for Model-Based Engineering of Embedded Systems. In: MBEES 2014 – Zehnter Workshop Modellbasierte Entwicklung eingebetteter Systeme. Dagstuhl, 2014.

[Schneider and Trapp 2013] D. Schneider, M. Trapp: Conditional safety certification of open adaptive systems. ACM Transactions on Autonomous and Adaptive Systems (TAAS), Vol. 8, No. 2, 2013.

[Serediouk 2013] V. Serediouk: UML-basierte Runtime Verification. Diploma Thesis, Humboldt-Universität Berlin, 2013.

[Schröck et al. 2013] S. Schröck, F. Zimmer, T. Holm, A. Fay, T. Jäger: Principles, viewpoints and effect links in the engineering of automated plants. In: Proceedings of IECON 2013 – 39th Annual Conference of the IEEE Industrial Electronics Society. Vienna, 2013.

[Schröck et al. 2014] S. Schröck, A. Fay, T. Jäger: Concept for the systematic reuse of functional units within the engineering of automated plants. In: Automation 2014: Fünfzehntes Branchentreff der Mess- und Automatisierungstechnik, Baden-Baden, 2014.

[Schröck et al. 2014] S. Schröck, F. Zimmer, A. Fay, T. Jäger: Konzept zur funktionsorientierten systematischen Wiederverwendung im Engineering automatisierter Anlagen der Prozessindustrie. In: Entwurf komplexer Automatisierungssysteme (EKA): Tagungsband der dreizehnten. Tagung, Magdeburg, 2014.

T

[Tenbergen et al. 2013] B. Tenbergen, P. Bohn, T. Weyer: Ein strukturierter Ansatz zur Ableitung methodenspezifischer UML/SysML-Profile am Beispiel des SPES 2020 Requirements. In: Workshopband Software Engineering 2013, GI-Edition, Lecture Notes in Informatics (LNI), Dritter Workshop zur Zukunft der Entwicklung softwareintensiver eingebetteter Systeme (ENVISION2020), 2013, pp. 235 – 244.

[Tenbergen et al. 2014] B. Tenbergen, A. C. Sturm, T. Weyer: A Hazard Taxonomy for Embedded and Cyber-Physical Systems. In: Proceedings of the 1st International Workshop on Emerging Ideas and Trends in Engineering of Cyber-Physical Systems (EITEC '14), 2014.

[Tenbergen et al. 2015] B. Tenbergen, T. Weyer, K. Pohl: Supporting the Validation of Adequacy in Requirements-based Hazard Mitigations. In: Proceedings of 21th International Working Conference on Requirements Engineering – Foundation for Software Quality (REFSQ 2015), Lecture Notes in Computer Science, 9013, Springer, 2015, pp. 17 – 32.

[Trapp et al. 2013] M. Trapp, D. Schneider, P. Liggesmeyer: A Safety Roadmap to Cyber-Physical Systems. In: Perspectives on the Future of Software Engineering, Spinter, 2013, pp. 81 – 94.

[Trapp et al. 2013] M. Trapp, J. C. Fabre, P. Queré: CARS: Critical Automotive Systems. In: Workshop-Proceedings of 2nd Workshop CARS: Robustness and Safety at Safecomp 2013, LAAS-CNRS, 2013.

[Trapp and Schneider 2013] M. Trapp, D. Schneider: Safety Assurance of Open Adaptive Systems – A Survey. In Models@run.time. Foundations, Applications, and Roadmaps, Springer, 2014, pp. 279 – 318.

V

[Vetrò and Diebold 2014] A. Vetrò and P. Diebold: Three-level taxonomy of technology transfer mediums. 2014.

[Vetrò et al. 2014] A. Vetrò, P. Diebold, F. Grigoleit, W. Böhm: An Evidence-based Strategy for Evaluating Project Success based on Technology Transfer. International Conference on Evaluation and Assessment in Software Engineering (EASE), 2014.

[Vetrò et al. 2015] A. Vetrò, W. Böhm and M. Torchiano: On the Benefits and Barriers when adopting Software Modelling and Model Driven Techniques – An external, differentiated replication. In: Empirical Software Engineering and Measurement (ESEM), 2015.

[Vetrò et al. 2015] A. Vetrò, P. Diebold, F. Grigoleit, W. Böhm: A Strategy for Evaluating Project Success based on Technology Transfer. In: Proceedings of Software Engineering & Management 2015. Lecture Notes in Informatics, GI Edition, 2015.

[Voss et al. 2014] S. Voss, J. Eder, F. Hölzl: Design Space Exploration and its Visualization in AUTO-FOCUS3. Vierter Workshop zur Zukunft der Entwicklung softwareintensiver eingebetteter Systeme (ENVISION2020), 2014.

W

[Wartenberg 2013] F. Wartenberg: Model-Based Test Design of Efficient Test Suites for Software Product Lines. Diploma Thesis, Humboldt-Universität Berlin, 2013.

[Weber 2014] R. Weber, E. Thaden, S. Henkler, J. Höfflinger, S. Prochnow: Design Space Exploration for an industrial Lane-Keeping-Support Case Study. DATE Conference – University Boot, 2014.

[Weißleder] S. Weißleder, H. Schlingloff: An Evaluation of Model-Based Testing in Embedded Applications. In: Proceedings of ICST 2014 – 7th IEEE International Conference on Software Testing, Verification and Validation. Cleveland, Ohio, 2014.

[Wehrstedt et al. 2012] J. C. Wehrstedt, C. Leuxner, R. Rosen: Towards Seamless Model-Based Design of Complex Systems with Early Consideration of Automation Software. Mechatronics. Linz, 2012.

[Weißleder et al. 2012] S. Weißleder, T. Girlich, J. Krause: Automatic Traceability from Tests to Requirements by Requirements – Based Refinement. In: FORMS/FORMAT 2012, 9th International Symposium on Formal Methods, Braunschweig, 2012.

[Weyer 2013] T. Weyer: Der Status Quo im Requirements Engineering – Herausforderungen und Lösungen durch modellbasierte Techniken. In: Tagungsband des 6. Embedded Software Engineering Kongress, Elektronikpraxis, Sindelfingen, 2013.

Z

[Zimmermann et al. 2015] W. Zimmermann, W. Böhm, C. Grelck, R. Heinrich, R. Jung, M. Konersmann, A. Schlaefer, E. Schmieders, S. Schupp, B. T. y Widemann, T. Weyer (Hrsg.): Gemeinsamer Tagungsband der Workshops der Tagung Software Engineering 2015. Dresden, CEUR Workshop Proceedings, 1337, 2015.

E – Index

A

Application domain
- Automation 4, 7
- Automotive 4, 7, 12
- Avionics 4, 7
- Energy 4
- Healthcare 4
- Rail 4
- Robotics 4

Artifact quality assessment framework 269

Automation
- Domain 217, 267, 291, 295
- Example 19
- Industry 265

Automotive
- Adaptive cruise control 148
- Case study 183
- Domain 189, 199, 200, 267, 291, 295, 297
- Engineering 119
- Example 13, 14, 181
- Industry 132, 198, 265
- Safety integrity level 200
- System cluster 12, 14, 48, 51, 88, 96, 134, 202

Avionics
- Collision avoidance system 148, 297
- Domain 217, 228, 267, 291, 295, 297
- Industry 265
- System 199, 228

B

Building blocks 31, 123, 140, 169, 193, 227

C

Case study 294, 295, 297, 302

Automation 19
Automotive 14, 183
Beach well 40
Industrial 155, 263, 290, 291

Challenges
- Early validation 83
- Embedded system development 4
- Engineering 7, 19, 22
- Modular safety assurance 190
- Optimal deployment 162
- System function networks 131

Context 5, 49, 55, 57, 68, 72, 82, 131
- Analysis 228
- Behavioral operational 56, 102, 117, 197
- Diagram 37, 72, 73
- Documentation 49, 195
- Entities 225
- Functional operational 55, 102, 135, 196
- Functions 134
- Model 33, 58, 85, 86, 99, 103, 116, 119, 257
- Modeling 198
- Modeling concepts 49
- Modeling framework 31, 46, 50, 58, 66, 85, 100, 134, 192, 194, 257
- of knowledge 51, 87, 193, 194, 201
- Operational 24, 54, 70, 85, 87, 99, 188
- Organizational 268
- Physical 114, 116
- Process 268
- Runtime 98
- Safety 191, 193
- Social 268
- Structural operational 54, 101, 117, 196
- Technical 176, 268
- Technology 268
- Variability 225

© Springer International Publishing AG 2016

K. Pohl et al. (eds.), *Advanced Model-Based Engineering of Embedded Systems*,
DOI 10.1007/978-3-319-48003-9

Continuous model-driven engineering 141

D

Degree of Granularity 34, 249
Deployment 133, 144, 159, 295
    Constraints 162
Domain knowledge model 87
DSE framework 163
DSE techniques 171

E

Engineering
    Artifacts 52, 80
    Challenges 7, 19, 22, 303
    Model-based 3
    Path 35
    Process 247
Example
    Automation 19
    Automotive 14

F

Functional design 141

G

Goal question metric 270
Goal structuring notation 209
Granularity
    Degree of 34, 249

H

Hazard analysis 188

M

Methodological framework 163
Modular safety assessment 295
Modular safety assurance 185
    Activities 188

O

Open Safety Model 191, 208

P

Pivot model 277
Process building block framework 37

R

Requirements quality assessment framework ... 84

S

Safety 187, 189
    Case 193, 204
    Contracts 207
    Goals 209
    Requirement decomposition pattern 205
    Requirements 204
Safety framework 190
SPES engineering methodology 5
SPES modeling framework 56
    Application 77
    Degree of Granularity 34
    Extension 117, 132, 134, 166, 218
    Viewpoints 32
SPES XT
    Key contributions 8
SPES XT modeling framework 29
    Context modeling framework 31
    Evaluation 288
    Extensions 31, 41, 46, 63
    Process building block framework 31
    Systems Engineering Extensions 32
SPES XT tool platform 273
System function networks 129

T

Technology transfer 261

V

Validation 59, 80, 86, 133, 146, 149

Early 83, 295

   Techniques 88

Variability 18

   Assessment 237

   Exchange language 232

   Management 232

   Perspective 224

Variant management 216, 282, 295

Verification..59, 114, 133

Viewpoint..32

   Functional 33, 252

   Logical... 33, 254

   Requirements...................................... 33, 164, 250

   Technical 33, 255

Printed in the United States
By Bookmasters